Introduction to Gastrointestinal Diseases Vol. 1

Jakub Fichna
Editor

Introduction to Gastrointestinal Diseases Vol. 1

Springer

Editor
Jakub Fichna
Department of Biochemistry
Medical University of Lodz
Lodz
Poland

ISBN 978-3-319-49015-1 ISBN 978-3-319-49016-8 (eBook)
https://doi.org/10.1007/978-3-319-49016-8

Library of Congress Control Number: 2016955335

This Springer imprint is published by the registered company Springer Nature Switzerland AG
The registered company address is: Gewerbestrasse 11, 6330 Cham, Switzerland

Preface

"We are what we eat" is a common saying, yet not always understood and applied, and not many of us know that the delivery of proper nutrients is necessary for maintenance of homeostasis and functioning of the entire body. On the other hand, only recently the gastrointestinal (GI) tract has emerged as a crucial system, intertwining structurally and functionally with the central and peripheral nervous as well as immune systems, and several others. Therefore, not only diet, but also lifestyle and surrounding environment may influence the GI tract and related systems.

In the recent years, functional and inflammatory diseases of the GI tract have been taking their toll and the number of their cases is significantly increasing, what triggers the need for extensive medical care. In case of irritable bowel syndrome (IBS), its management constitutes 25–50 % of the entire gastroenterology outpatients workload. As for inflammatory bowel disease (IBD), the incidence of Crohn's disease (CD) is estimated at 5 per 100,000 people per year and the prevalence is 40–50 per 100,000 people in the Western and Northern Europe. The incidence and prevalence of CD are maintained at a stable level in developed countries, whereas in developing areas these rates are constantly growing. The incidence rate of ulcerative colitis (UC) is about 10 per 100,000 people.

Furthermore, the average age of onset of the GI tract diseases becomes a serious concern. Approximately 25 % of IBD cases are diagnosed in the first two decades of patient's life, especially in childhood (age 13–18). The highest incidence of UC occurs already between the ages of 20 and 40.

Alarmingly, only the minority of patients (e.g., one-third in the case of IBS) seek advice from a general practitioner; many do not consider their symptoms serious enough to consult the doctor and usually seek different treatment modalities, not always acceptable from the medical point of view or efficient. Moreover, only 20 % of patients—when they do not respond to conventional treatment—are referred to see a gastroenterologist. Finally, IBS and IBD patients often look for medical information from the Internet, which does not necessarily provide the same quality of knowledge as official brochures, books, or medical professionals.

Through this book, we hope to change the current situation for the patient and for the doctor. The book has been prepared by professionals in basic and clinical gastroenterology, therefore the information provided is up to date and of highest quality. Moreover, we focus on both, the patient and the doctor. We hope that through this book we will encourage a new approach to the management of the GI tract diseases not only by educating the patient and the doctor, but also showing that the collaboration between them is beneficial for better diagnosis and cure.

Lodz, Poland Jakub Fichna

In collaboration with Natalia Fabisiak, Paula Mosińska and Maciej Sałaga.

Contents

Introduction

Jakub Fichna

Abstract The chapter focuses on composition and function of the gastrointestinal tract, primarily in physiological conditions. The information provided in the chapter will give basis for understanding the malfunction of the digestive system that leads to diseases described further in the book.

Keywords Gastrointestinal tract • Small and large bowel • Digestion • Gastrointestinal motility • Water and electrolyte transport

The gastrointestinal (GI) tract (or *alimentary canal*) is often regarded as a simple tube that allows us turning meals into body muscle (or fat) and that from time to time—may cause some trouble, whether before a stressful interview or once we eat too many sour cherries. In fact, there is much more "magic" in the functioning of the GI tract and the digestive system plays a much more important role than expected.

Five basic functions of the GI tract that refer to food and nutrient processing are ingestion, propulsion, digestion (chemical and mechanical), absorption and elimination. However, the GI tract cross talks with several other systems, thus its role in immune defense, water and electrolyte homeostasis or—at the time of disease—delivery of therapeutics cannot be forgotten. The GI tract is sometimes, and not necessarily wrongfully, compared to skin that separates us from the outside environment, yet allows communication with external stimuli and translation of this signaling so that it can be understood by the body.

Food goes first to the mouth, where accessory digestive organs (teeth, tongue, salivary glands along with palatal surfaces) start the process of ingestion and digestion, first through mechanical processing and breaking down (through tearing, chewing, mashing, and crushing) and then mixing with saliva (for detailed information on food digestion in the GI tract please see Table 1). Saliva not only moistens the food matter, but also contains amylase, an enzyme that helps digest

Table 1 Food digestion in the gastrointestinal tract

Nutrient	Organ	Enzyme	End product
Carbohydrates (Polysaccharides)	Mouth	Salivary amylase	Oligosaccharides
	Duodenum and small intestine	Pancreatic amylase	Disaccharides and monosaccharides
	Small intestine	Brush border enzymes	Monosaccharides
Proteins	Stomach	Pepsin in presence of HCl	Polypeptides and oligopeptides and amino acids
	Duodenum and small intestine	Trypsin and chymotrypsin	Oligopeptides and dipeptides and amino acids
	Duodenum and small intestine	Carboxypeptidase	Oligopeptides and dipeptides and amino acids
	Small intestine	Aminopeptidase and dipeptidase	Amino acids
Lipids	Duodenum and small intestine	Pancreatic lipase in presence of bile salts	

carbohydrates, as well as immunoglobulins, lysozyme and other bacteriostatics and antibiotics. Last but not least, let's not forget the tongue and its role in taste sensing through taste buds.

The processed food (or *bolus*) may now proceed to esophagus, a 20–25 cm long tube (all measures for an average adult male) that transports it rapidly to the stomach. The passage through esophagus is facilitated by mucus secretion and peristaltic movement of its walls due to visceral muscles located within. A small flap called epiglottis is unequivocally important at this stage—located in the pharynx, it prevents from food entering the trachea by covering and closing its entrance.

Stomach (1 L when empty, 2–4 L when full) is the first organ of the digestive tract where the food can be stored for some time, allowing further mixing and digestion. The chemicals excreted by gastric glands in the stomach wall (for GI tract histology please see Box 1; for GI tract control please see Box 2) are hydrochloric acid (HCl, an inorganic acid) and pepsinogen (a precursor molecule to an enzyme pepsin) that allow digestion of proteins. Of note, enteroendocrine cells in the stomach wall secrete specific hormones that influence the GI tract functioning, like gastrin, histamine, endorphins, serotonin, cholecystokinin, and somatostatin. Further processing of carbohydrates under the influence of salivary amylase may also be possible in the stomach, yet harsh acidic conditions do not promote high activity of the enzyme.

Box 1. Histology of the Gastrointestinal Tract
The gastrointestinal tract wall has four layers

1. Mucosa (secretes mucus, digestive enzymes, and hormones; absorbs nutrients; protects deeper parts of the wall from damaging conditions and the entire system against ingested infectious and toxic factors)
   - Epithelium
   - Lamina propria
   - Muscularis mucosae
2. Submucosa (connective tissue with large blood vessels, lymphatics, and nerves branching into the mucosa and muscularis externa; contains an enteric nervous plexus called the submucosal plexus)
3. Muscularis Externa (responsible for peristaltic contractions and segmentation movements; forms sphincters; contains myenteric plexus between two muscle layers)
   - Longitudinal muscle
   - Circular muscle
4. Serosa (for intraperitoneal and retroperitoneal organs) or Adventitia (for esophagus and retroperitoneal organs)
   - Epithelium
   - Connective tissue.

Box 2. Neuronal Control of the Gastrointestinal Tract
Local (involves enteric nervous system):

- Submucosal plexus (controls glands and mucosal muscle)
- Myenteric plexus (controls GI motility)

General (involves central nervous system):

- Parasympathetic (enhances GI motility and secretion)
- Sympathetic (decreases GI motility and secretion).

Food (in a form of chyme) now propagates to the small intestine, composed of the duodenum, the jejunum, and the ileum. The duodenum, measuring 20–25 cm plays multiple roles in food processing: it receives chyme from the stomach, due to higher pH (=alkaline conditions) it neutralizes gastric acid before it enters further into the small intestine, and receives "juices" (digestive secretions) from the pancreas and the liver. The pancreas, which is regarded as an accessory digestive organ, secretes (1) into the pancreatic juice (by exocrine cells)—enzymes necessary for food digestion, such as amylase (breaks down carbohydrates), lipase (breaks down lipids), proteases (break proteins), and peptidases (break peptides into amino acids); (2) into the bloodstream (by endocrine cells of the pancreatic islets)—insulin and glucagon, which participate in glucose homeostasis. Pancreatic secretion may be regulated neuronally (through parasympathetic nerves) and hormonally (through gut hormones cholecystokinin and secretin). Liver on the other hand is like a large laboratory in our body, involved in (1) direct or indirect regulation of hundreds of biochemical and hematological processes and (2) bile production, which is an important emulsifier indispensable for dispersing lipids into smaller droplets, whose surface is available for pancreatic lipase. It needs to be mentioned that bile, before it gets to the duodenum, is stored by the gallbladder and released upon a specific gut hormone, cholecystokinin. Secretin, another gut hormone, stimulates bile secretion.

From the duodenum, the chyme passes to the jejunum (2.5 m long) and the ileum (3.5 m long), where further mixing, chemical digestion and nutrient absorption to lymph and blood occurs. Lining of the small intestine wall has finger-like projections called villi, covered in microvilli which further increase surface area for absorption. The chyme in the small intestine is kept moist by intestinal secretions, what allows action of digestive enzymes which require aqueous environment, and it propagates through peristaltic contractions and segmentation movements, which are independent of the brain control. It has been estimated that it takes on average 5 h for the chyme to pass from the duodenum to the distal part of the ileum.

Next step is the large intestine (or large bowel, 1.5 m), which divides into the cecum (receives material from the ileum which is then stored and compacted), the colon (ascending—transverse—descending—sigmoid) and the rectum. In the large intestine, the undigested contents is formed into and temporarily stored as feces (later expelled through defecation through the anus), and reabsorption of excess of water and bile salts (which are transported with blood back to the liver) occurs. Similarly to the small intestine, the movement of the contents is due to peristaltic waves ("along the length") and segmentation movements ("churning"), yet much slower up to transverse colon, what allows water absorption. Defecation is triggered by distension of rectal wall, which is detected through specific stretch receptors. In physiological conditions, feces (on average 80–220 g/day) contain unused or indigestible food, mucus, and dead cells that used to line the GI tract.

Importantly, the large intestine is colonized by several bacterial strains, which produce vitamins (K, B_5, biotin) that are also absorbed within the colon. They also process bilirubin (end product of red blood cell breakdown) to yellow-to-brown colorants of the fecal matter, as well as peptides and carbohydrates to intestinal gases.

Part I
Irritable Bowel Syndrome and Functional Gastrointestinal Diseases

Chapter 1
Current Theories for Development of Irritable Bowel Syndrome

Paula Mosińska and Julia Krajewska

List of Abbreviations

5-HT	Serotonin; 5-hydroxytryptamine
ACTH	Adrenocorticotropic hormone
BGA	Brain-gut axis
CNS	Central nervous system
CRF	Corticotrophin releasing factor
CRP	C-reactive protein
ECs	Enterochromaffin cells
ENS	Enteric nervous system
FODMAPs	Fermentable oligosaccharides, disaccharides, monosaccharides and polyols
GI	Gastrointestinal tract
GRID2IP	Glutamate receptor ionotropic delta 2 interacting protein; delphilin
HPA	Hypothalamic-pituitary-adrenal
HTR3	5-hydroxytryptamine receptor 3; serotonin receptor 3
IBD	Inflammatory bowel disease
IBS	Irritable bowel syndrome
IBS-C	Constipation-predominant IBS
IBS-D	Diarrhea-predominant IBS
IELs	Intraepithelial lymphocytes
IFN-γ	Interferon-γ
GPCR	G protein-coupled receptor
GR	Glucocorticoid receptor
IL	Interleukin
KDELR2	KDEL endoplasmic reticulum protein retention receptor 2

P. Mosińska (✉) · J. Krajewska (✉)
Department of Biochemistry, Faculty of Medicine, Medical University of Lodz,
Mazowiecka 6/8, 92-215, Lodz, Poland
e-mail: paula.mosinska@gmail.com

J. Krajewska
e-mail: krajewska.julia@gmail.com

J. Fichna (ed.), *Introduction to gastrointestinal diseases Vol. 1*,
DOI 10.1007/978-3-319-49016-8_1

M1	Proinflammatory classically activated macrophage
M2	Antiinflammatory alternatively activated macrophage
miRs	MicroRNAs
MCs	Mast cells
NO	Nitric oxide
PAI-1	Plasminogen activator-1
PAR-2	Protease-activated receptor-2
PI-IBS	Post infectious IBS
PGs	Prostaglandins
PM	Particulate matter
PVN	Paraventricular nucleus
ROS	Reactive oxygen species
SCFA	Short chain fatty acids
SERT	5-HT reuptake transporter protein
SNP	Single nucleotide polymorphisms
TGF-β	Transforming growth factor β
Th	T helper cells
TJs	Tight junctions
TL1A	TNF-like ligand 1A
TLRs	Toll-like receptors
TNF-α	Tumor necrosis factor- α
TNFSF15	TNF ligand superfamily member 15 gene
TRPV1	Transient receptor potential vanilloid type 1
WAT	White adipose tissue
WT	Wild type
ZO	Zonula occludens

Irritable bowel syndrome (IBS) is a chronic, relapsing functional bowel disorder associated with altered gastrointestinal (GI) motility, secretion and sensation. It is the most commonly diagnosed functional GI condition, with the highest prevalence in Western, rather than Eastern countries. North America and Europe remain the leading regions of IBS incidence, with about 10–20 % cases (up to approximately 25 % in some studies) in both adolescents and adults. Nevertheless, a rise in disease toll has recently been observed in South China (11.5 %) and Korea (6.6 %). Overall, the prevalence of IBS within the population is estimated between 10 and 25 %; the discrepancy often results from different criteria used for diagnosis of IBS, as well as the influence of other fluctuating factors, including the population evaluated and their access to health care, the cultural impact and the scanty data of inclusion or exclusion of comorbid disorders.

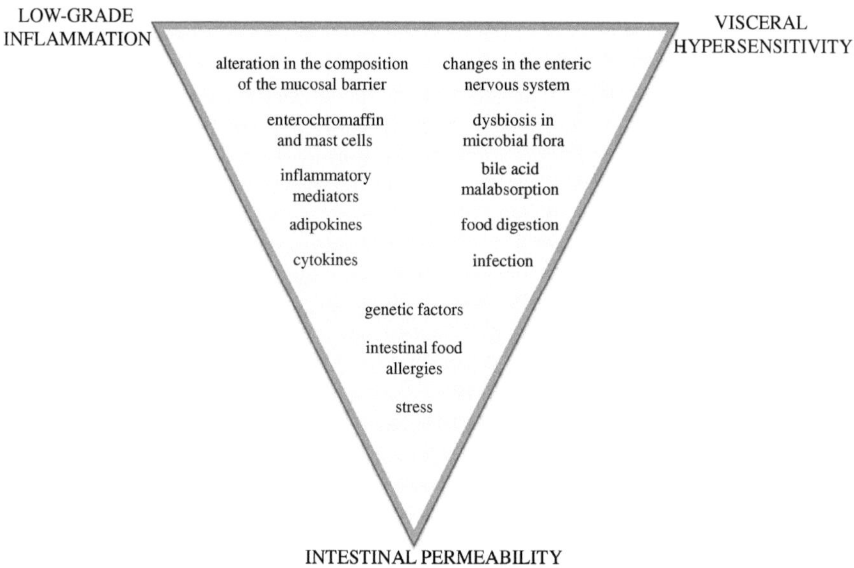

Fig. 1.1 Correlation between intestinal permeability, low-grade inflammation, and visceral hypersensitivity—factors involved in IBS etiopathology

Etiology and pathophysiology of IBS is complex and not well understood, but it is most likely multifactorial. Mechanisms that have an established role in initiation or progress of IBS include increased mucosal permeability, visceral hypersensitivity, intestinal mucosa activation, and an interplay between luminal factors, the epithelial barrier, and the mucosal immune system (Fig. 1.1).

The probability of IBS development is higher among those individuals with a biological relative with IBS. However, some reports claim that the heredity of IBS can occur as a result of similar psychosocial, behavioral, psychological or environmental factors rather than genetic causes.

Considerable evidence links the central nervous system (CNS) mechanisms with symptoms experienced by IBS patients. Accordingly, bidirectional interactions in the brain-gut axis (BGA) and changes in the number and type of released neurotransmitters can affect endocrine, autonomic, immune and motor functions. Very often psychological co-morbidities e.g. stress, anxiety, depression and phobic disorder additionally influence and exacerbate symptom perception in IBS patients. Moreover, the association between serum tumor necrosis factor (TNF)-α and anxiety, as well as mucosal MCs infiltration and fatigue, provide the proof of the relevance of psychological factors and BGA in IBS pathophysiology [1].

Along with abnormal microbiota, food allergies, previous infections, or bile acid malabsorption, environmental pollution has recently gained much attention.

1.1 Increased Mucosal Permeability

Intestinal epithelium plays a pivotal role as a barrier that selectively regulates paracellular permeability and transport, and therefore prevents access by potentially harmful pathogens and their toxins from the intestinal lumen. One of the major components of the intestinal mucosal barrier are the intracellular tight junctions (TJs), which—along with occludin and claudins—comprise a complex system that, by interacting with specific zonula occludens (ZO)-1, ZO-2 and ZO-3 proteins regulates permeability to small molecules and electrolytes (Fig. 1.2). An abnormal TJs structure or function alters intestinal permeability and may contribute to worsening of IBS symptoms [2]. In line, in vivo studies as well as assays on mucosal biopsies of the small and large intestines from diarrhea-predominant-IBS (IBS-D) and constipation-predominant-IBS (IBS-C) patients revealed an augmented intestinal permeability and decreased mRNA expression of ZO-1 and occludin, compared with healthy controls [3]. In contrast, Camilleri et al. [4] reported an increased expression of occludin in the intestinal mucosa in patients with IBS-C; hence, to date it is still unclear whether the expression of occludin is more likely associated with diarrhea or constipation IBS. Several in vivo studies in mice also showed increased intestinal permeability upon administration of colonic supernatants from all IBS subtypes and fecal supernatants from IBS-D patients, which was positively correlated with somatic and visceral pain.

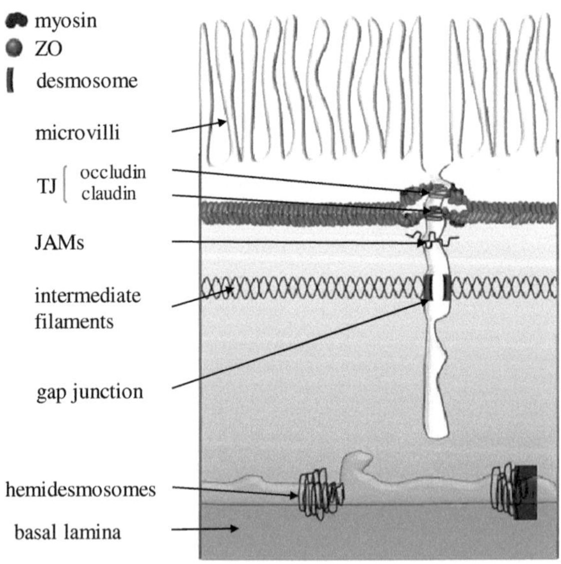

Fig. 1.2 The architecture of intestinal epithelia. *JAMs* junctional adhesion molecules, *TJ* tight junction, *ZO* zona occludens

1.1.1 Microbiota

It is possible that changes in the intestinal microbiota and their ability to modify systemic immune response directly by release of soluble peptides and toxins, and indirectly via the induction of cytokines, may have much in common with impairment of TJs complexity [5, 6]. Intestinal bacteria can directly modulate the epithelial barrier function, for example through the secretion of fermented dietary fiber products, such as short chain fatty acids (SCFA), e.g. acetate, propionate or butyrate. Butyrate, in particular, was seen to accelerate the expression of TJ proteins in vitro, induce the intestinal angiogenesis in vivo, and finally participate in differentiation and growth of enterocytes. Acetate and propionate, in turn, are substrates for glucose and lipid synthesis mainly in the liver. The indirect action of intestinal microbes depends on the type of cytokines released as they can weaken [e.g. TNF-α, interferon-γ (IFN-γ)], or strengthen [e.g. transforming growth factor β (TGF-β), interleukin (IL)-10] the intestinal barrier [2] (see below).

Quantitative and/or qualitative changes in the microbial gut population trigger bloating, one of the symptoms commonly experienced by women. Also worth mentioning, impaired intestinal barrier and increased permeability have also been documented in patients with infectious gastroenteritis and individuals with post infectious IBS (PI-IBS); after acute infection the intestinal permeability rises two to threefold.

1.1.2 Other Factors Influencing Intestinal Permeability

More recently, depletion of glutamine in the lower GI tract in IBS patients is considered as an important contributor for increased permeability of the intestinal barrier. Its deficiency affects claudin-1 expression in the colonic mucosa, and reduces proliferation of enterocytes, which causes further epithelial atrophy. Glutamine action seems to be dependent on basal expression of TJ proteins [7].

Inflammatory mediators or antimicrobial peptides also participate in the modulation of the function of TJs. As a result of immune activation, a higher infiltration of cytokines and proinflammatory mediators released by mast cells (MCs) is observed in the colonic mucosa of IBS patients [8]. Proinflammatory cytokines found in IBS colonic samples, such as IL-4, IL-6, IL-12, IL-1β, TNF-α and IFN-γ contribute to occludin damage, which further impairs the TJ complexity [9]. On the other hand, many cytokines released during inflammation exhibit the antiinflammatory action, e.g. IL-10 or TGF-β by preserving intestinal disruption and inflammatory development. Importantly, the levels of these mediators are significantly decreased in the colon and rectum of IBS patients.

Noteworthy, the activation of MCs, rather than their number is presumed to be associated with the alterations observed in the intestinal mucosal barrier (For review see [10]).

1.2 Visceral Hypersensitivity

Alterations within the intestinal epithelium affect sensineural processing and change both motor and sensory activities of the gut, contributing to symptom progression in IBS patients. However, extra-epithelial factors may also contribute to visceral hypersensitivity and abdominal pain.

MCs serve as a contributor in various intestinal disorders, including IBS, inflammatory bowel disease (IBD) or food allergy. The density of MCs in both small and large intestine is increased in IBS patients, compared to healthy individuals, but their distribution within the lower GI tract varies—some reports demonstrate higher accumulation in the small rather than large intestine, while others show the opposite. Deviations from the number of MCs appear independently of gender and the subtype of IBS; nonetheless, the presence of MCs per se does not always imply a pathogenic importance, unless they are activated.

Due to close proximity of MCs to enteric nerves and varicosities, MCs activation increases enteric neuron excitability and enhances potential firing of extrinsic sensory neurons in a MC-derived, mediator dependent manner. Multiple factors such as bacteria, viruses, parasites, toxins or endogenous peptides may activate MCs, prompting their degranulation and subsequent release of various mediators from cytoplasmic granules, particularly histamine, chymase, tryptase and cytokines, or induce synthesis of leukotriene C_4, platelet activating factor or prostaglandins (PGs). There is a growing appreciation for the hypothesis that MCs activation positively correlates with intestinal symptoms, such as abdominal pain, discomfort, bloating, changes in bowel habits, and psychological symptoms including depression and fatigue (for review see [11]).

Several studies reported a high concentration of histamine, proteases and tryptase in IBS mucosa, which excite and sensitize sensory nerves. Interestingly, histamine directly increases sensory response through H_1 receptor-mediated mechanism. The blockade of H_1 receptor results in a decreased response of mesenteric afferents, and inactivation of proteases in the mucosal supernatants of IBS patients [12]. Tryptase, in turn, is abundantly expressed in MC granules and is considered as a marker for their activation. It possibly activates spinal afferent terminals and enhances intestinal permeability through the protease-activated receptor-2 (PAR-2) on enterocytes, which subsequently causes long-lasting neuronal hyperexcitability and redistribution of TJs allowing the intraepithelial passage of macromolecules [13]. Interestingly, infusion of mucosal supernatants from IBS patients into the colon of rodents i.e. mice or rats, induces visceral hyperalgesia and allodynia. Both effects are reversed by PAR-2 antagonist and serine protease inhibitors. No hyperalgesia was documented in PAR-2—deficient mice [14], what supports its role in sensory activation.

Serum samples from IBS sufferers contain elevated levels of proinflammatory IL-1β, IL-6, IL-8 and TNF-α, which directly affect neuronal activity and alter intestinal contractility, absorption and/or secretion. Notably, breakdown of the mucosal barrier by proinflammatory cytokines allows foreign particles to invade the intestinal barrier and induce the immune response in the submucosal and myenteric neuronal plexi (for review see: [15]). Interestingly, IBS-D patients have higher accumulation of MCs within the small intestine, which causes defects in apical junctional complex integrity, by increasing the spaces between epithelial cells, and generally contributes to exacerbation of clinical symptoms, such as diarrhea and pain [16].

There is substantial literature providing evidence for the association between the frequency and severity of abdominal pain experienced by IBS patients and the presence of MCs within the gut. More recently, it has been shown that an allergic background worsens IBS symptoms by enhancing the infiltration of MCs into the cecum and rectum mucosa, and promoting the secretion of soluble factors responsible for paracellular permeability. Patients with food allergy are more susceptible to develop symptoms typical for IBS-D, rather than IBS-C, which probably results from a different level of tryptase released by MCs or even other pathways involved in MC activation. This notion warrants further investigation [17].

Apart from immune cells, enterochromaffin (EC) cells, whose serum concentration is increased especially in IBS-D subtype, are responsible for the modulation of nerve activity in the epithelium of lower GI tract by the release of serotonin (5-hydroxytryptamine, 5-HT). Because 5-HT participates in multiple GI functions, including vasodilation, peristalsis, electrolyte secretion and absorption, and in pain perception, it is reasonable to think that the symptoms observed in IBS can stem from alterations in serotonergic signaling. The highest level of 5-HT is found in the amygdala, integral to the emotional responses and visceral stimulation, which indicates that BGA also plays a critical role in signal transmission. The 5-HT reuptake transporter protein (SERT) regulates the action of 5-HT receptor within the GI tract by maintaining transmitter homeostasis and terminating the transmission. Worth noting, in SERT knockout mice, the level of 5-HT augments in an uncontrolled manner which aggravates the inflammatory response [18]. Several studies demonstrate that gene expression of SERT is downregulated in the colon and rectum of IBS patients, which is associated with increased 5-HT mucosal availability, augmented reflex activity and luminal hypersecretion that frequently result in diarrhea-like symptoms [19]. Moreover, the colonic 5-HT release correlates with MC infiltration and thereby can drive abdominal pain, irrespective of IBS subtype. Overall, IBS-D patients have considerably higher number of EC cells in comparison with IBS-C patients; nonetheless, some reports reveal no differences in EC quantity between IBS subtypes.

These data indicate that any disturbances in expression and/or content of 5-HT receptors, as well as changes in SERT activity impact sensorimotor function and thereby affect severity of abdominal pain/discomfort perceived by IBS patients. Additionally, the therapeutic efficacy of drugs affecting 5-HT receptors also supports the contribution of 5-HT in IBS pathophysiology (for more information see: Clinical treatment).

More recently, the upregulation and sensitization of receptors located on the peripheral nerve terminals of nociceptors are thought to imply visceral hypersensitivity. Notably, a high number of mucosal sensory transient receptor potential cation channel subfamily V type 1 (TRPV1) has been identified in the distal colon of IBS patients. The number of TRPV1 fibres was up to threefold higher in colonic biopsies obtained from IBS-C and IBS-D patients, than in control individuals. Moreover, a positive correlation between TRPV1 staining intensity, visceral pain and low-grade inflammatory infiltration was observed [20, 21].

1.3 Low-Grade Inflammation

Although IBS is not generally considered as an inflammatory disease, several lines of evidence indicate that immune and inflammatory mechanisms contribute to its pathophysiology. A low-grade inflammatory response, which changes the enteric neuromuscular and sensory nerve function, has been observed in the GI tract in patients with either "conventional", as well as PI-IBS. Among various factors that can favor IBS symptoms in these patients, an increased number of immunocytes, especially lamina propria T cells, toll-like receptors (TLRs), intraepithelial lymphocytes (IELs), and—as previously mentioned—mucosal MCs, have been outlined.

Type 1 T-helper (Th)-1 and Th-2 cells are able to restrain one another and produce cytokines to maintain a balanced immune response. Dependent on the segment of the intestine and concomitant infectious disease, different expression level of Th-1- and Th-2-derived cytokines and various quantity of immunocytes occur in IBS patients. The peripheral blood obtained from IBS-D patients is abundant in Th-1 derived cytokines, such as proinflammatory IFN-γ, IL-2 and IL-12, and lower number of Th-2—derived cytokines, e.g. IL-4. Higher levels of TNF-α, IL-1β, IL-6 and LPS-induced IL-6 were also observed mostly in the subset of IBS-D. Similarly, in the intestinal mucosa of PI-IBS patients, where the ratio of Th-1 to Th-2 is shifted towards Th-1-derived cytokines, an augmented level of IFN-γ and IL-4 and significantly decreased level of antiinflammatory IL-10 was documented [22]. The imbalance between Th-1 and Th-2 is possibly driven by the infection, which changes the epithelial permeability and prompts subsequent response against microbiotic agents. Although some discrepancies exist in the number of cytokines secreted either by Th-1 or Th-2 cells, and detected in various studies of different IBS subtypes, the single nucleotide polymorphisms (SNP) in the major proinflammatory TNF-α, secreted by Th-1, and impaired production of antiinflammatory IL-10 have been detected, and implied in the mechanism of low-grade inflammation in IBS [23].

TNF-α is a particularly important cytokine produced primarily by macrophages and monocytes, which participates in the pathogenesis of many inflammatory diseases e.g. IBD. TNF-α circulates within the body where it activates MCs and neutrophils, modulates the action of vascular endothelial cells, exhibits tumoricidal activity and orchestrates the cytokine cascade in various immune states. TNF-α, along with IL-6, also control the hypothalamic-pituitary-adrenal (HPA) axis via corticotrophin-releasing hormone. Hence, many studies quantified the level of this prominent inflammatory mediator in patients with IBS.

A significant proportion of IBS patients have an elevated TNF-α in the serum/plasma and stool samples, when compared to healthy individuals [24, 25]; however, while analyzing the IBS subgroups, elevated TNF-α is most prevalent in IBS-D [26]. The level of TNF-α was also increased in peripheral-blood cells among IBS patients, what may influence the symptom perception in the gut mucosa via activation of sensory nerve pathways. Interestingly, changes in the serum level of TNF-α and other cytokines e.g. IL-5, Il-13 are associated with higher anxiety and/or depression [1, 27] (for more information please see Visceral hypersensitivity). The majority of studies highlighted the importance of TNF-α in the pathogenesis of IBS; nonetheless, there are also several studies in which the level of TNF-α was similar as in the healthy individuals. The variability of findings may result from the small sample size, variations in laboratory assays used in the study or heterogeneity of subject data—as up to now IBS is generally symptom-based.

White adipose tissue (WAT), besides its ability to respond to afferent signals from the CNS and hormones, also expresses and secretes both pro- and antiinflammatory adipokines, such as cytokines, chemokines, and hormone—like factors, which participate in a variety of physiological or pathophysiological processes (Fig. 1.3). The crosstalk between the inflamed intestine and the surrounding mesenteric adipose tissue is currently investigated in view of the etiopathology of

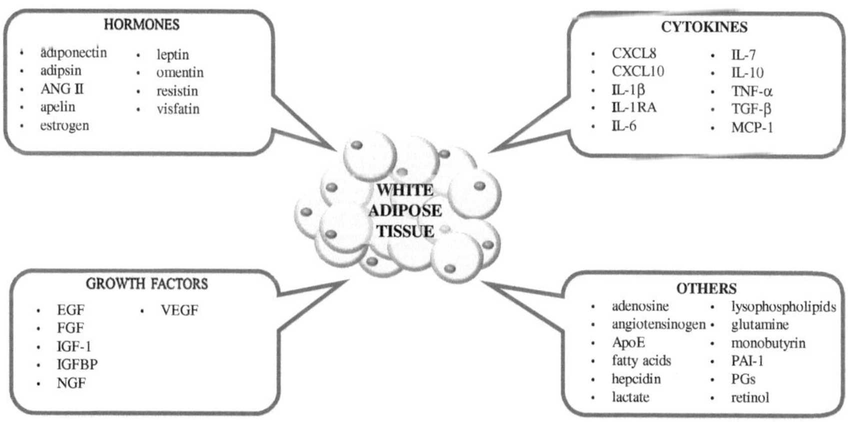

Fig. 1.3 The complex function of white adipose tissue (WAT) in the synthesis and secretion of adipokines and lipids. *ANG II* angiotensin II, *ApoE* apolipoprotein E, *CXCL8* C-X-C motif chemokine 8, *CXCL10* C-X-C motif chemokine 10; *interleukin 8* ligand 8, *EGF* epidermal growth factor, *FGF* fibroblast growth factor, *IGF-1* insulin-like growth factor 1, *IGFBP* insulin-like growth factor-binding protein. *IL-1* interleukin 1β, *IL-1RA* interleukin 1receptor antagonist, *IL-6* interleukin 6, *IL-7* interleukin 7, *IL-10* interleukin 10, *MCP-1* monocyte chemoattractant protein 1, *NGF* nerve growth factor, *PAI-1* plasminogen activator inhibitor 1, *PGs* prostaglandins, *TGF-β* transforming growth factor β, *TNF-α* tumor necrosis factor α, *VEGF* vascular endothelial growth factor

IBS [28]. Consequently, resistin, leptin, TNF-α, plasminogen activator-1 (PAI-1), IL-6 and angiotensinogen are proinflammatory mediators, whereas adiponectin exerts antinflammatory activity. An overexpression of adipokines, particularly adiponectin, leptin and resistin were documented in mesenteric adipose tissue, while in IBS-D subjects the level of resistin and leptin were increased but adiponectin decreased [29].

The antiinflammatory properties of adiponectin encompass the inhibition of proinflammatory IL-6 and simultaneously induction of the antiinflammatory cytokines IL-10 and IL-1 [30]. Moreover, adiponectin can modulate macrophage phenotype by switching it from the proinflammatory classically activated macrophage (M1) to an antiinflammatory alternatively activated macrophage (M2) [31]. Accordingly, adiponectin knockout mice have elevated expression of M1 markers, including TNF-α and IL-6 in macrophages and stromal vascular fraction (SVFs), as compared with wild type (WT) mice. Administration of adiponectin to WT mice contributes to higher expression of M2-related genes and IL-10 [31]. Nevertheless, not all studies confirm the antinflammatory effect of adiponectin. It was observed that an increased adiponectin level occurs in inflamed, rather than non-inflamed mesenteric adipose tissue, which indicates that adiponectin exerts proinflammatory effects on colonic epithelial cells. In chronic autoimmune or inflammatory diseases the level of adiponectin seems to be elevated. The inconsistency in the reports presenting adiponectin as either proinflammatory or antinflammatory adipokine, may depend on the excess of adipose tissue during inflammation. It is hypothesized that in chronic and autoimmune diseases, such as rheumatoid arthritis, type 1 diabetes or IBD, the increase in adiponectin may be responsible for inflammation-induced catabolic responses. However, an unambiguous determination of what kind of process adiponectin is responsible for is still not possible.

Unlike adiponectin, resistin exerts proinflammatory effect by inducing expression of vascular endothelial adhesion molecules, what leads to infiltration of leukocytes into the site of immune reaction. Exogenous treatment of mice with recombinant human resistin regulates the release of proinflammatory cytokines from macrophages and adipocytes, including TNF-α, IL-6 and IL-1, which confirms its role in the process of inflammation.

Among a wide range of actions of leptin, e.g. control of the feeding behavior, modulation of the satiety, thermogenesis and lipid and glucose metabolism, the activation and modulation of various cytokines seem particularly important in the pathogenesis of immune and inflammatory disorders. This adipokine stimulates proliferation of naïve T-helper lymphocytes, controls the expression of IL-1, IL-2, IL-6 and TNF-α by T-cells [32, 33], promotes generation of NO and proinflammatory cytokines by macrophages, and contributes to the release of reactive oxygen species (ROS) from neutrophils. The concentration of leptin is higher in IBS-D versus control subjects, indicating its presumable involvement in IBS pathophysiology [29].

Along with a fair number of inflammatory cytokines which participate in the development of low-grade inflammation, proinflammatory adipokines may also indirectly affect the function of sensory innervations of the gut, and thus promote visceral pain.

Of interest, higher endotoxin levels have been observed in a subset of IBS patients, in comparison to healthy control. Gram-negative bacteria-derived endotoxins stimulate various proinflammatory mediators such as TNF-α, and acute phase reactants, including C-reactive protein (CRP), which prompt intestinal mucosal damage and further augment the extent of intestinal permeability and inflammation [34].

Collectively, it seems that increased permeability of the intestinal barrier, low-grade inflammatory response and altered 5-HT signaling may account for local sensitization and enhanced visceral pain observed in IBS patients (Fig. 1.1).

1.4 Genetic Predisposition

The observation of familial aggregation of IBS has pointed to the idea that genetic background contributes to IBS development. It has been demonstrated that relatives of IBS sufferers are at increased risk of IBS. Twin studies presented consistent results, showing higher concordance rate for IBS between monozygotic than dizygotic twins. Although environmental factors could be responsible for the familial aggregation of IBS, the fact that no association was found between spouses indicates the role of genes. Several genetic factors that seem to be associated with IBS have been recognized [35].

Polymorphisms related to serotonergic system have been found, in agreement with the known function of 5-HT in the pathogenesis of IBS. In particular, a polymorphism in the regulatory region of the SERT gene results in a long (L) or short (S) allele. Homozygous S/S genotype was found to be related to IBS-C and IBS-D in many studies, although some other did not confirm these results. This genotype may result in decreased expression of the gene and, consequently, reduced 5-HT reuptake [35, 36]. In another line of research, single nucleotide polymorphisms (SNPs) associated with IBS-D were found in HTR3 genes encoding serotonin type 3 receptors [37].

Genetic variants connected with neuronal function influence visceral sensitivity. Namely, IBS-associated polymorphisms in genes encoding voltage-gated sodium channel, neurexophilin 1, adrenergic receptors, members of the opioid and cannabinoid receptors, catechol-O-methyltransferase, brain-derived neurotrophic factor and fatty acid amide hydrolase have been reported [35].

Impairment of the immune system function is among the pathophysiological factors of IBS (for more information see: Low-grade inflammation). SNPs associated with IBS were recognized for TNF ligand superfamily member 15 gene (TNFSF15), suggesting it may have a role in immune modulation in IBS. TNFSF15 gene product, TNF-like ligand 1A (TL1A) is expressed in immune cells and participates in the regulation of the inflammatory response, particularly in interactions with pathogens and commensal bacteria in the gut. [38] Polymorphisms in genes associated with intestinal barrier function have also been reported [35].

The results mentioned above were obtained by studies conducted on candidate genes. Additionally, hypothesis-free genome-wide association study revealed two more genes associated with IBS: KDEL endoplasmic reticulum protein retention receptor 2 (KDELR2) and glutamate receptor ionotropic delta 2 interacting protein (GRID2IP) in the 7p22.1 region. However, there is little understanding regarding the function of these genes in IBS pathogenesis. KDELR2 is hypothesized to act as a mediator in the activity of bacterial toxins in the intestines. GRID2IP, also known as delphilin, is expressed in synapses in the brain, and interacts with glutamate receptor δ2 and monocarboxylate transporter 2. Either of these interactions could play a role in IBS, through faulty activity of glutamatergic signaling or alteration of interaction between host and intestinal microbiota [39].

Apart from genetic factors, epigenetics may also be relevant in IBS pathophysiology. Although the research in this area has not been extensive, there are some reports which correlate specific epigenetic factors with IBS. Firstly, stress response can be mediated by epigenetic changes. One study reported changes in methylation patterns of glucocorticoid receptor (GR) and corticotrophin releasing factor (CRF) genes in amygdalae of rats presenting visceral hypersensitivity in response to chronic psychological stress. Furthermore, administration of histone deacetylase inhibitor attenuated this stress-evoked hyperalgesia. These results suggest that central epigenetic mechanisms are implicated in stress-induced visceral pain [40].

Secondly, the expression of several microRNAs (miRs) is altered in IBS. A subset of patients with IBS-D had increased expression of miR-29a and miR-29b [41]. MiR-29a targets glutamine synthetase. Since glutamine is essential for proper structure of the intestinal mucosa, its decreased level leads to an increased intestinal permeability [42]. Moreover, the increase in miR-29a and miR-29b downregulated the expression of claudin-1 (a tight junction protein) and nuclear factor κB-repressing factor (an inhibitor of NF-κB activity), which was associated with excessive intestinal permeability [41]. In another study, a decrease in miR-199 expression and a consequent increase in transient receptor potential vanilloid type 1 (TRPV1) expression was associated with increased visceral sensitivity. TRPV1, a cation channel in peripheral afferent sensory neurons, modulates pain perception and mechanosensation in animal models of IBS related to stress and enteritis [43]. In a preliminary study, the increased expression of miR-150, and miR-342-3p, most likely connected with inflammatory processes and pain modulation, was also associated with IBS [44].

Conflicting results obtained by various groups make it difficult to draw general conclusions regarding the role of genes in the pathogenesis of IBS. It should also be noted that a discovered statistical association does not necessarily imply causal relationship. The GENes in Irritable Bowel Syndrome Research Network EURope (GENIEUR) Action may help obtain more consistent and reliable outcomes and bring new light onto the role of genetic factors in the pathogenesis of IBS [35].

1.5 Brain-Gut Axis

BGA is a bidirectional communication system which consists of the CNS, the enteric nervous system (ENS) within the gut wall, and the link between them: the neural, endocrine as well as neuroimmune pathways (Fig. 1.4) [45]. Although ENS can control the basic gut functions, such as motility, secretion, absorption, and local blood flow, the connection is required to maintain homeostatic balance of the organism. The brain influences the gut, while the signals from the GI tract are transmitted to the brain. In this way, proper GI function, such as food intake, digestion and control of the bowel movements, is maintained. Dysregulation of BGA is considered as one of the causes of functional GI syndromes, including IBS [46].

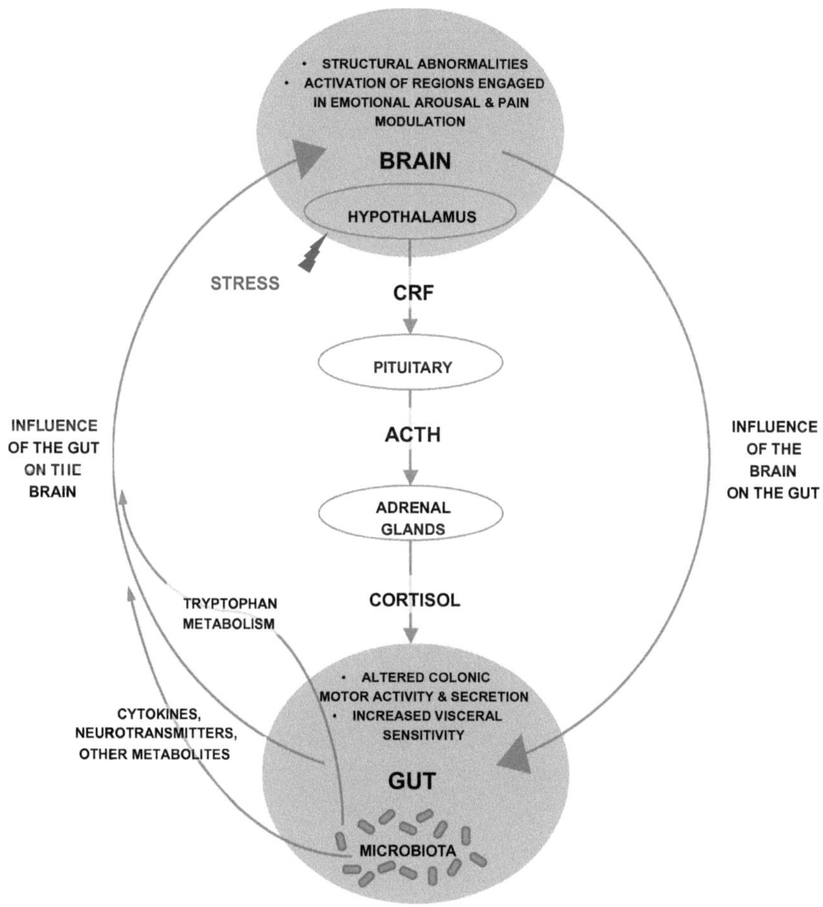

Fig. 1.4 Brain-gut axis in IBS

Modern imaging techniques allowed to identify structural changes in IBS patients as compared to the healthy individuals, including thinning in the anterior midcingulate and insular cortex, structures related to interoception, as well as regional increases and decreases in gray matter density, predominantly in areas involved in attention, modulation of emotion and pain. White matter abnormalities were also observed in areas connected with pain and visceral perception [46]. Moreover, functional neuroimaging techniques allowed to observe differences in brain activation in IBS. In particular, regions engaged in emotional arousal (anterior cingulate cortex, amygdala) and midbrain cluster, involved in endogenous pain modulation, were more activated. On the contrary, activation in regions responsible for visceral sensation was similar in patients with IBS and healthy individuals, although its extent and specific location of foci varied. Additionally, medial and lateral prefrontal cortex, associated with cognitive modulation of pain and emotion, showed greater activity in controls [47].

Signaling system initiated by CRF secretion—the first stage in activation of hypothalamus-pituitary-adrenal (HPA) axis—is another component implicated in the pathogenesis of IBS. CRF, released by the paraventricular nucleus (PVN) of the hypothalamus, is a key element of the physiological response to stress. In response to CRF, the pituitary gland secretes adrenocorticotropic hormone (ACTH), which in turn stimulates cortisol release from the adrenal glands. CRF mediates the response to stress through CRF1 receptor, which belongs to the G protein-coupled receptor (GPCR) family. Stimulation of CRF1 induces colonic motor activity and secretion as well as visceral sensitivity, which suggests its involvement in IBS development. Activation of CRF1 receptors both in the brain (centrally) and in the GI tract (peripherally) evokes IBS symptoms, while CRF receptor antagonists neutralize the influence of stress on the gut [48, 49].

Indeed, IBS is known to be related to stress and psychiatric disorders. The prevalence of psychiatric and psychological disorders among IBS patients is high, typically estimated between 40 and 60 %, and their presence corresponds with the severity of IBS symptoms. The most common comorbid psychiatric disorders include mood disorders (depression and dysthymia), anxiety disorders (including panic disorder, generalized anxiety disorder and post-traumatic stress disorder) and somatoform disorders, although other, e.g. drug or alcohol abuse, have also been reported. Other psychosocial factors associated with IBS include sexual and emotional abuse history, stressful life events (e.g. divorce, unemployment, death of a close relative), specific personality traits (neuroticism and alexithymia), hypochondriasis and maladaptive coping strategies. There are several hypotheses that explain how stress influences IBS development, the most widely accepted being the modulation of BGA and alteration of the immune response [50, 51].

1.6 Alterations in Gut Microbiota

Gut microbiome is a fundamental component of the physiology of its host organism, essential for the maintenance of homeostasis. Microbiota play an important role in the function of GI system, motility, absorption and secretion as well as permeability of the intestinal wall, as represented by the commonly known fact that imbalance can result in diarrhea (e.g. in the course of infection or antibiotic treatment). The understanding of the role of microbiota is still limited. It remains unclear to what extent the variations in microbiota composition are a cause of the disorder, and how much IBS symptoms and dietary restrictions introduced by the patients influence the microbiome. Moreover, the shifts in microbiota may occur in some, but not all patients suffering from IBS, which might hinder the elucidation of their role in the disorder [45, 52].

Several arguments support the role of microbiota in the pathogenesis of IBS. Firstly, multiple studies demonstrate alterations in microbiota composition in IBS patients in comparison to healthy individuals. However, there is a significant variability between the results and a consistent microbial profile cannot be determined. The results may vary due to differences in sampling and experimental techniques, as well as various IBS diagnostic criteria and patient populations. The consistent finding seems to be decreased diversity and instability of the composition of the gut microbiota in IBS subjects [45, 52]. Several groups confirmed the increased abundance of *Firmicutes* and decreased proportion of *Bacteroidetes* [45, 53]. Secondly, visceral hypersensitivity can be observed when the microbiome is changed due to infection or antibiotic treatment. On the other hand, certain probiotic strains can ameliorate IBS symptoms. In particular, *Bifidobacterium* sp. and *Lactobacillus* sp. have proven effective in clinical trials [52]. Notably, antibiotics may increase the risk of IBS, but in some cases, e.g. rifaximin, they may also have a positive influence [45]. Another significant indication of the role of microbiome comes from the studies of germ-free animals, as they present abnormal function of the immune system and differences in GI system, but also impaired stress response and social-cognitive deficits [45, 52]. These findings exemplify the influence of gut microbiome on the CNS and psychosocial functions of the host organism.

The microbiota-gut-brain axis, a modification of the BGA, describes the interaction between the gut microbiome and the CNS of the host. This communication takes place through neural, hormonal and immunological pathways. Although various mechanisms could explain these interactions, tryptophan metabolism currently attracts particular attention. Gut bacteria regulate the peripheral concentrations of tryptophan, which is crucial for 5-HT synthesis in the CNS. 5-HT is a critical mediator both in CNS and ENS. Additionally, increased intestinal permeability that occurs in IBS may facilitate the release of bacterial metabolites, e.g. inflammatory cytokines, into the bloodstream and influence the host [45].

The majority of IBS sufferers associate the onset or aggravation of their symptoms with the meals and identify food items that worsen their condition, therefore they exclude particular products from their diet. As certain dietary products are digested, the metabolites of microbiota may participate in IBS manifestations generation. Grains, vegetables, dairy products, fat, spices, coffee and alcohol are among the diet components frequently related to IBS symptoms. In the course of digestion, the fermentation of carbohydrates primarily produces SCFA, hydrogen and carbon dioxide. FODMAPs, which include fermentable oligosaccharides, disaccharides, monosaccharides and polyols, constitute a group of particular interest with regard to IBS. Their fermentation may lead to excessive gas production and osmotic effect, causing abdominal pain, flatulence and diarrhea. As proteins are fermented, detrimental products that are formed include ammonia, hydrogen sulfide, organic acids, heterocyclic amides, phenolic and indolic compounds. Fatty products and proteins enhance bile acids release, which may affect microbiota composition and GI functions (for additional information see: Patient's guide) [53].

1.7 Environmental Pollution

Recently, environmental factors, including microbial contamination, radiation, air pollution and stress, have been proposed to play a role in the development of IBS.

Pathogenic microbes cause enteric infections, which may result in PI-IBS. The precise mechanism of the disorder development has not been recognized, however, dysfunctions in the mucosal wall and alterations in intestinal microbiome may be of importance.

Polluted air contains gases, such as carbon dioxide, ozone, NO, volatile organic compounds, and particulate matter (PM). People exposed to air pollution ingest PM, which may lead to inflammatory response and damaging of the colon mucosa. Air pollution may also influence the microbiota composition [54].

Radiation exposure may influence the immune system, possibly leading to intestinal inflammation, which, in turn, may result in IBS development.

Finally, life in modern societies is associated with increased susceptibility to stress, which is one of the psychosocial factors implicated in IBS pathogenesis [55].

1.8 Summary

Although the pathogenesis of IBS remains unclear, several factors that contribute to the development of the disorder have been identified. Increased intestinal permeability described in IBS sufferers results from impaired mucosal barrier. Visceral hypersensitivity might be a consequence of altered sensineural processing, mast cells activation, alterations in serotonergic signaling, as well as upregulation and sensitization of nociceptors. Moreover, low-grade inflammation, which modifies enteric nerve function, occurs in IBS. Next, increase in pro-inflammatory cytokines, adipokines secreted by WAT and high endotoxin levels are reported in IBS patients.

Genetic variants, which have a role in the pathogenesis, are related to serotonergic system, visceral sensitivity and modulation of immune system. Alterations in epigenetic factors may also be of importance. Dysregulation of the BGA, such as structural changes within the brain and impairment of HPA axis are considered as one of the causes of the disease, which is in agreement with the known association of IBS with stress and psychiatric disorders. Change in the composition and diversity of the gut microbiota is also connected with the condition. Finally, pollution of the environment may be a factor which contributes to IBS development. All in all, pathogenesis of IBS is multifactorial, and many components still remain to be elucidated.

References

1. Kindt S, Van Oudenhove L, Broekaert D et al (2009) Immune dysfunction in patients with functional gastrointestinal disorders. Neurogastroenterol Motil 21:389–398. doi:10.1111/j.1365-2982.2008.01220.x
2. Camilleri M, Madsen K, Spiller R et al (2012) Intestinal barrier function in health and gastrointestinal disease. Neurogastroenterol Motil 24:503–512. doi:10.1111/j.1365-2982.2012.01921.x
3. Crişan I-M, Dumitraşcu DL (2014) Irritable bowel syndrome: peripheral mechanisms and therapeutic implications. Clujul Med 87:73–79. doi:10.15386/cjmed-269
4. Camilleri M, Carlson P, Acosta A, Busciglio I (2015) Colonic mucosal gene expression and genotype in irritable bowel syndrome patients with normal or elevated fecal bile acid excretion. Am J Physiol Gastrointest Liver Physiol 309:G10–G20. doi:10.1152/ajpgi.00080.2015
5. Distrutti E, Monaldi L, Di Gastro- SC et al (2016) Gut microbiota role in irritable bowel syndrome: new therapeutic strategies. World J Gastroenterol Febr 21:2219–2241. doi:10.3748/wjg.v22.i7.2219
6. Jandhyala SM, Talukdar R, Subramanyam C et al (2015) Role of the normal gut microbiota. World J Gastroenterol 21:8787–8803. doi:10.3748/wjg.v21.i29.8787
7. Bertrand J, Ghouzali I, Guérin C et al (2015) Glutamine restores tight junction protein claudin-1 expression in colonic mucosa of patients with diarrhea-predominant irritable bowel syndrome. J Parenter Enteral Nutr. doi:10.1177/0148607115587330
8. Barbara G, Zecchi L, Barbaro R et al (2012) Mucosal permeability and immune activation as potential therapeutic targets of probiotics in irritable bowel syndrome. J Clin Gastroenterol 46 (Suppl):S52–S55. doi:10.1097/MCG.0b013e318264e918
9. Piche T (2014) Tight junctions and IBS—the link between epithelial permeability, low-grade inflammation, and symptom generation? Neurogastroenterol Motil 26:296–302. doi:10.1111/nmo.12315
10. Matricon J, Meleine M, Gelot A et al (2012) Review article: associations between immune activation, intestinal permeability and the irritable bowel syndrome. Aliment Pharmacol Ther 36:1009–1031. doi:10.1111/apt.12080
11. Wouters MM, Vicario M, Santos J (2015a) The role of mast cells in functional GI disorders. Gut 65:155–168. doi:10.1136/gutjnl-2015-309151
12. Barbara G, Wang B, Stanghellini V et al (2007) Mast cell-dependent excitation of visceral-nociceptive sensory neurons in irritable bowel syndrome. Gastroenterology 132:26–37. doi:10.1053/j.gastro.2006.11.039

13. Li M, Zhang L, Lu B et al (2015) Role of dendritic cell-mediated abnormal immune response in visceral hypersensitivity. Int J Clin Exp Med 8:13243–13250
14. Greenwood-Van Meerveld B, Moloney RD, Johnson AC, Vicario M (2016) Mechanisms of stress-induced visceral pain: implications in irritable bowel syndrome. J Neuroendocrinol. doi:10.1111/jne.12361
15. Wouters MM, Vicario M, Santos J (2015b) The role of mast cells in functional GI disorders. Gut 155–168. doi:10.1136/gutjnl-2015-309151
16. Martínez C, Lobo B, Pigrau M et al (2013) Diarrhoea-predominant irritable bowel syndrome: an organic disorder with structural abnormalities in the jejunal epithelial barrier. Gut 62:1160–1168. doi:10.1136/gutjnl-2012-302093
17. Vivinus-Nebot M, Dainese R, Anty R et al (2012) Combination of allergic factors can worsen diarrheic irritable bowel syndrome: role of barrier defects and mast cells. Am J Gastroenterol 107:75–81. doi:10.1038/ajg.2011.315
18. Bischoff SC, Barbara G, Buurman W et al (2014) Intestinal permeability—a new target for disease prevention and therapy. BMC Gastroenterol 14:189. doi:10.1186/s12876-014-0189-7
19. Coates MD, Mahoney CR, Linden DR et al (2004) Molecular defects in mucosal serotonin content and decreased serotonin reuptake transporter in ulcerative colitis and irritable bowel syndrome. Gastroenterology 126:1657–1664. doi:10.1053/j.gastro.2004.03.013
20. Akbar a, Yiangou Y, Facer P, et al (2008) Increased capsaicin receptor TRPV1-expressing sensory fibres in irritable bowel syndrome and their correlation with abdominal pain. Gut 57:923–929. doi:10.1136/gut.2007.138982
21. Wouters MM, Balemans D, Van Wanrooy S et al (2016) Histamine receptor H1-mediated sensitization of TRPV1 mediates visceral hypersensitivity and symptoms in patients with irritable bowel syndrome. gastroenterology. doi:10.1053/j.gastro.2015.12.034
22. Chen J, Zhang Y, Deng Z (2012) Imbalanced shift of cytokine expression between T helper 1 and T helper 2 (Th1/Th2) in intestinal mucosa of patients with post-infectious irritable bowel syndrome. BMC Gastroenterol 12:91. doi:10.1186/1471-230X-12-91
23. Bashashati M, Rezaei N, Shafieyoun A et al (2014) Cytokine imbalance in irritable bowel syndrome: a systematic review and meta-analysis. Neurogastroenterol Motil 26:1036–1048. doi:10.1111/nmo.12358
24. Darkoh C, Comer L, Zewdie G et al (2014) Chemotactic chemokines are important in the pathogenesis of irritable bowel syndrome. PLoS One 9:e93144. doi:10.1371/journal.pone.0093144
25. Seyedmirzaee S, Hayatbakhsh MM, Ahmadi B et al (2016) Serum immune biomarkers in irritable bowel syndrome. Clin Res Hepatol Gastroenterol. doi:10.1016/j.clinre.2015.12.013
26. van der Veek PPJ, van den Berg M, de Kroon YE et al (2005) Role of tumor necrosis factor-alpha and interleukin-10 gene polymorphisms in irritable bowel syndrome. Am J Gastroenterol 100:2510–2516. doi:10.1111/j.1572-0241.2005.00257.x
27. McKernan DP, Gaszner G, Quigley EM et al (2011) Altered peripheral toll-like receptor responses in the irritable bowel syndrome. Aliment Pharmacol Ther 33:1045–1052. doi:10.1111/j.1365-2036.2011.04624.x
28. Lee CG, Lee JK, Kang Y-S et al (2015) Visceral abdominal obesity is associated with an increased risk of irritable bowel syndrome. Am J Gastroenterol 110:310–319. doi:10.1038/ajg.2014.422
29. Russo F, Chimienti G, Clemente C et al (2013) Adipokine profile in celiac patients: differences in comparison with patients suffering from diarrhea-predominant IBS and healthy subjects. Scand J Gastroenterol 48:1377–1385. doi:10.3109/00365521.2013.845907
30. Caselli C, D'Amico A, Cabiati M et al (2014) Back to the heart: the protective role of adiponectin. Pharmacol Res 82:9–20. doi:10.1016/j.phrs.2014.03.003
31. Lovren F, Pan Y, Quan A et al (2010) Adiponectin primes human monocytes into alternative anti-inflammatory M2 macrophages. Am J Physiol Heart Circ Physiol 299:H656–H663. doi:10.1152/ajpheart.00115.2010

32. Waluga M, Hartleb M, Boryczka G et al (2014) Serum adipokines in inflammatory bowel disease. World J Gastroenterol 20:6912–6917. doi:10.3748/wjg.v20.i22.6912
33. Yarandi SS, Hebbar G, Sauer CG et al (2011) Diverse roles of leptin in the gastrointestinal tract: modulation of motility, absorption, growth, and inflammation. Nutrition 27:269–275. doi:10.1016/j.nut.2010.07.004
34. Yoon JY, Park SJ, Cheon JH (2014) Effect of colostrum on the symptoms and mucosal permeability in patients with irritable bowel syndrome: a randomized placebo-controlled study. Intest Res 12:80–82. doi:10.5217/ir.2014.12.1.80
35. Gazouli M, Wouters MM, Kapur-Pojskić L et al (2016) Lessons learned—resolving the enigma of genetic factors in IBS. Nat Rev Gastroenterol Hepatol 13:77–87. doi:10.1038/nrgastro.2015.206
36. Makker J, Chilimuri S, Bella JN (2015) Genetic epidemiology of irritable bowel syndrome. World J Gastroenterol 21:11353–11361. doi:10.3748/wjg.v21.i40.11353
37. Gu QY, Zhang J, Feng YC et al (2015) Association of genetic polymorphisms in HTR3A and HTR3E with diarrhea predominant irritable bowel syndrome. Int J Clin Exp Med 8:4581–4585
38. Zucchelli M, Camilleri M, Andreasson AN et al (2011) Association of TNFSF15 polymorphism with irritable bowel syndrome. Gut 60:1671–1677
39. Ek WE, Reznichenko A, Ripke S et al (2014) Exploring the genetics of irritable bowel syndrome : a GWA study in the general population and replication in multinational case-control cohorts. Gut 1–9. doi:10.1136/gutjnl-2014-307997
40. Tran L, Chaloner A, Sawalha AH, Greenwood Van-Meerveld B (2013) Importance of epigenetic mechanisms in visceral pain induced by chronic water avoidance stress. Psychoneuroendocrinology 38:898–906
41. Zhou Q, Costinean S, Croce CM et al (2015) MicroRNA 29 targets nuclear factor-κB-repressing factor and claudin 1 to increase intestinal permeability. Gastroenterology 148:158–169. doi:10.1053/j.gastro.2014.09.037
42. Zhou Q, Souba WW, Croce CM, Verne GN (2010) MicroRNA-29a regulates intestinal membrane permeability in patients with irritable bowel syndrome. Gut 59:775–784
43. Zhou Q, Yang L, Larson S, et al (2015b) Decreased miR-199 augments visceral pain in patients with IBS through translational upregulation of TRPV1. Gut 1–9. doi:10.1136/gutjnl-2013-306464
44. Fourie NH, Peace RM, Abey SK et al (2014) Elevated circulating miR-150 and miR-342-3p in patients with irritable bowel syndrome. Exp Mol Pathol 96:422–425
45. Kennedy PJ, Cryan JF, Dinan TG, Clarke G (2014) Irritable bowel syndrome: a microbiome-gut-brain axis disorder? World J Gastroenterol 20:14105–14125. doi:10.3748/wjg.v20.i39.14105
46. Fichna J, Storr MA (2012) Brain-gut interactions in IBS. Front Pharmacol 3 JUL:1–12. doi:10.3389/fphar.2012.00127
47. Tillisch K, Mayer EA, Labus JS (2011) Quantitative meta-analysis identifies brain regions activated during rectal distension in irritable bowel syndrome. Gastroenterology 140:91–100. doi:10.1053/j.gastro.2010.07.053
48. Fukudo S (2007) Role of corticotropin-releasing hormone in irritable bowel syndrome and intestinal inflammation. J Gastroenterol 42(1):48–51. doi:10.1007/s00535-006-1942-7
49. Taché Y, Bonaz B (2007) Corticotropin-releasing factor receptors and stress-related alterations of gut motor function. J Clin Invest 117:33–40. doi:10.1172/JCI30085
50. Qin H-Y, Cheng C-W, Tang X-D, Bian Z-X (2014) Impact of psychological stress on irritable bowel syndrome. World J Gastroenterol 20:14126–14131. doi:10.3748/wjg.v20.i39.14126
51. Surdea-Blaga T, Băban A, Dumitrascu DL (2012) Psychosocial determinants of irritable bowel syndrome. World J Gastroenterol 18:616–626. doi:10.3748/wjg.v18.i7.616
52. De Palma G, Collins SM, Bercik P (2014) The microbiota-gut-brain axis in functional gastrointestinal disorders. Gut Microbes 5:419–429. doi:10.4161/gmic.29417

53. Rajilić-Stojanović M, Jonkers DM, Salonen A et al (2015) Intestinal microbiota and diet in IBS: causes, consequences, or epiphenomena? Am J Gastroenterol 110:278–287. doi:10.1038/ajg.2014.427
54. Kish L, Hotte N, Kaplan GG et al (2013) Environmental particulate matter induces murine intestinal inflammatory responses and alters the gut microbiome. PLoS One 8
55. Marynowski M, Likońska A, Zatorski H, Fichna J (2015) Role of environmental pollution in irritable bowel syndrome. World J Gastroenterol 21:11371–11378. doi:10.3748/wjg.v21.i40.11371

Chapter 2
Clinical Features of Irritable Bowel Syndrome

Paula Mosińska and Maciej Sałaga

There is no specific reliable biomarker for IBS, therefore the condition is defined predominantly by symptoms; however, because of their tremendous magnitude, e.g. recurrent abdominal pain and/or discomfort, altered stool consistency and frequency, distention or bloating the diagnosis of IBS is troublesome (see Box 2.1). The occurrence of these ailments may further prompt dehydration, sleep deprivation, anxiety or lethargy and lead to time off work, social awareness and contribute to overall decrease in quality of life.

Box 2.1. Warning signs experienced by IBS patients that warrant consultation with GP

All people presenting susceptibility for IBS should be examined whether they have any "red flags" indicators such as unintentional and unexplained weight loss, progressive or unrelenting pain, family history of bowel or ovarian cancer, rectal bleeding or changes in bowel habits for more than 3 weeks, or 6 weeks in a person aged over 60 years. The presence of warning signs may point at a greater probability of disease occurrence, thus their exclusion should be done as quickly as possible.

The proper diagnosis can be made if the above-mentioned factors will be accompanied with at least two of the following:

- changing in stool passage (straining, urgency, incomplete evacuation)
- bloating, distention, tension, hardness
- worsening of symptoms after food intake
- passage of mucus.

P. Mosińska (✉) · M. Sałaga
Department of Biochemistry, Faculty of Medicine, Medical University of Lodz, Mazowiecka 6/8, 92-215 Lodz, Poland
e-mail: paula.mosinska@gmail.com

J. Fichna (ed.), *Introduction to gastrointestinal diseases Vol. 1*,
DOI 10.1007/978-3-319-49016-8_2

Additionally, the occurrence of other symptoms such as nausea, backache or bladder symptoms may be used to support the diagnosis of IBS.

A detailed history of the abdominal pain and any associated symptoms e.g. the onset, duration, site, frequency and factors that can ease the pain are important characteristics in determining IBS.

Importantly, approximately 30 % of IBS-D patients experience loss of bowel control. For those patients, the symptoms impose a considerable constraint on everyday life—fear of an attack of diarrhea that limits their activities such as shopping, holidays or work.

2.1 Men Versus Women and IBS

Gender differences in IBS are well-established—the rate of IBS in general is 1.5–3-fold higher in women than men. Women are more sensitive and seek support from healthcare much more frequently. Moreover, they are able to depict their condition more precisely in comparison to men, which can have a reflection in current general rate.

The classic example of IBS patient is a young woman experiencing abdominal pain that is relieved by passage of multiple lose liquid stools. Her symptoms generally appear for more than 3 months and can be exacerbated by e.g. stress or diet intolerance. Worth mentioning, due to late luteal and early menses phases women frequently report nausea and other extraintestinal symptoms.

Overall, women are twice as likely as men to suffer from constipation-associated symptoms such as bloating, straining, abdominal distension and feelings of incomplete evacuation. Men tend towards diarrhea-associated symptoms, including watery stools, enhanced stool frequency, abdominal pain and bloating [8].

2.2 Abdominal Pain

Visceral hypersensitivity is present in all IBS subtypes albeit its intensity shifts from mild to severe.

It is generally accepted that patients with mild-to-moderate IBS frequently suffer from peripherally-generated symptoms, including intermittent, abdominal pain, which is relieved by defecation [5]. Increase in intestinal motility or visceral afferent firing can promote and amplify patient's ailments, and simultaneously increase psychological distress, which—along with psychosocial contributors—exacerbates symptom intensity. This vicious circle may significantly affect one's quality of life, thus the severity of IBS has to be understood from different points of view including peripheral signs as well as central dysregulation.

Without any doubt, abdominal pain is the most bothersome symptom which frequently serves as a predictor to seek medical care by IBS sufferers. Its intensity is augmented after food intake or during time of stress. Additionally, in contrast to other phases of menstrual cycle, abdominal and rectal pain, and bloating are worsened by menstruation.

2.3 Distention

A study, which comprised of 2259 IBS subjects demonstrated that patients with IBS-C are 14-fold more likely to suffer from distention, when compared to controls. The probable cause is the augmentation in bacterial overgrowth, due to slower colonic transit that in turn prompts colonic fermentation and increases distention with gas and stool. Vast majority of IBS patients are aware of their condition and many of them should easily describe in detail the anatomical localization and type of symptoms they experience.

2.4 Bloating

Abdominal fullness, pressure or a sensation of trapped gas is commonly experienced by IBS patients, more frequently among IBS-C than IBS-D. Sufferers report worsening of bloating especially after meals, with the tendency of this symptom to disappear overnight. Among females with IBS, bloating occurs commonly in the lower abdomen, perhaps owing to abnormalities of intestinal gas handling; whether the site of symptom is related to gynecological factors is still unknown.

In general, abdominal bloating positively correlates with the degree of abdominal distention solely in IBS-C, indicating different pathophysiology between subtypes of IBS [7].

2.5 Nausea

The study on 144 IBS patients revealed gender-related preponderance in the occurrence of nausea—about 2.5-fold higher incidence rate was observed in woman than men [1]. Additionally, in the study of 714 ROME I positive IBS patients, nausea was shown to be more common in premenopausal than postmenopausal women; however, this disparity might stem from the effect of the menstrual cycle, and be independent of the intake of oral contraceptives [6]. Of note, the variability in reproductive hormone levels during the menstrual cycle and changes in ovarian function at menopause impact visceral sensitivity and GI motility much more strongly among woman with IBS, rather than healthy individuals.

2.6 Psychological Disorders

Irrespective of the subtype, patients suffering from IBS tend to exhibit psychiatric and psychosocial disturbances, they are more vulnerable to stress and generally exhibit a high degree of abnormal illness behavior. Over half of all IBS patients report more severe somatic symptoms, including symptom-related fears, anxiety, depression or somatization [2, 3].

Worth mentioning, nearly half of all IBS cases coexist with other functional disorders such as chronic fatigue syndrome, chronic headache, temporomandibular joint dysfunction, fibromyalgia, or chronic pelvic or back pain.

2.7 Non-gastrointestinal Symptoms

The majority of IBS patients exert a range of extra-gastrointestinal symptoms, indicating the concomitant involvement of other non-GI organs, and suggest a generalized rearrangement of the central nervous system. The most common include: lethargy, headache, backache, urinary-associated symptoms (nocturia, urgency of micturition, incomplete bladder emptying, dyspareunia) [9].

2.8 Acute and Chronic Symptoms in IBS

The clinical course of IBS is chronic, although symptoms vary and oscillate not only between subtypes, but also within the same patient over time. At the beginning, symptoms may fluctuate over a short period of time i.e. weeks or months, but in many cases the severity tends to stabilize over 1–2 years follow-up. Some patients, especially those undiagnosed or those who do not implement their GPs' recommendations into their daily life, may develop new symptoms—not always specifically attributed to their IBS subtype—which may rise the possibility of occurrence of other organic pathology [4].

References

1. Anbardan SJ, Daryani NE, Fereshtehnejad S-M et al (2012) Gender role in irritable bowel syndrome: a comparison of irritable bowel syndrome module (ROME III) between male and female patients. J Neurogastroenterol Motil 18:70–77. doi:10.5056/jnm.2012.18.1.70
2. Burbige EJ (2010) Irritable bowel syndrome: diagnostic approaches in clinical practice. Clin Exp Gastroenterol 3:127–137. doi:10.2147/CEG.S12596
3. Canavan C, West J, Card T (2014) The epidemiology of irritable bowel syndrome. Clin Epidemiol 6:71–80. doi:10.2147/CLEP.S40245
4. Canavan C, West J, Card T (2015) Change in quality of life for patients with irritable bowel syndrome following referral to a gastroenterologist: a cohort study. PLoS ONE 10:1–13. doi:10.1371/journal.pone.0139389
5. Drossman DA, Chang L, Bellamy N, et al (2011) Severity in irritable bowel syndrome: a Rome Foundation Working Team report. Am J Gastroenterol 106:1749–1759; quiz 1760. doi:10.1038/ajg.2011.201
6. Heitkemper MM, Cain KC, Jarrett ME et al (2003) Symptoms across the menstrual cycle in women with irritable bowel syndrome. Am J Gastroenterol 98:420–430. doi:10.1111/j.1572-0241.2003.07233.x
7. Houghton LA, Lea R, Agrawal A, et al CLINICAL–ALIMENTARY TRACT Relationship of abdominal bloating to distention in irritable bowel syndrome and effect of bowel habit. doi: 10.1053/j.gastro.2006.07.015
8. Jiang X, Locke Iii GR, Choung RS, et al Prevalence and risk factors for abdominal bloating and visible distention: a population-based study. doi: 10.1136/gut.2007.142810
9. Spiller R, Aziz Q, Creed F et al (2007) Guidelines on the irritable bowel syndrome: mechanisms and practical management. Gut 56:1770–1798. doi:10.1136/gut.2007.119446

Chapter 3
Irritable Bowel Syndrome: Diagnosis

Marcin Włodarczyk and Aleksandra Sobolewska-Włodarczyk

Abstract Irritable bowel syndrome (IBS) is a chronic gastrointestinal disorder diagnosed on symptom-based criteria without inclusion of any objective parameter measurable by known diagnostic methods. Heterogeneity of patient's symptoms and overlapping with more serious organic diseases increase uncertainty for the physician's work and enhance the cost of confirming the diagnosis. In 2016 Rome IV criteria of functional disorders were published. These criteria are the basis to make a diagnosis in the daily work of medical doctors, especially general practitioners and gastroenterologists. Recent studies showed that in the future a combination of several new biomarkers could improve the diagnostic process of IBS. Among the studied biomarkers, most evidence is provided for fecal calprotectin. However, cut-off values for fecal calprotectin still have to be investigated prior to inclusion in the IBS diagnostic algorithm. In this chapter diagnosis criteria of IBS will be discussed.

According to the National Collaborating Centre for Nursing and Supportive Care (NICE) guidelines, prevalence of IBS is between 10 and 20 % worldwide, with women to men ratio 2:1 [1]. The need for a reliable and standard method to properly discriminate functional gastrointestinal disorders (FGIDs) has led to the development of symptom-based criteria by the Rome Foundation. Accordingly, diagnosis

M. Włodarczyk (✉) · A. Sobolewska-Włodarczyk
Department of Biochemistry, Faculty of Medicine, Medical University of Lodz, Mazowiecka 6/8, 92-215, Lodz, Poland
e-mail: dr.mwlodarczyk@gmail.com

A. Sobolewska-Włodarczyk
e-mail: olasobolewska1@poczta.onet.pl

M. Włodarczyk
Department of General and Colorectal Surgery, Faculty of Military Medicine, Medical University of Lodz, Hallera Sq 1, 90-647, Lodz, Poland

A. Sobolewska-Włodarczyk
Department of Gastroenterology, Faculty of Military Medicine, Medical University of Lodz, Hallera Sq 1, 90-647, Lodz, Poland

J. Fichna (ed.), *Introduction to gastrointestinal diseases Vol. 1*,
DOI 10.1007/978-3-319-49016-8_3

of IBS is established on clinical background with exclusion of "red flag" symptoms (age >50, rectal bleeding, anemia, short-term symptoms, and weight loss) [1]. Because there are usually no physical signs to definitively diagnose IBS, diagnosis is often a process of ruling out other conditions. To make the right identification, general practitioner or gastroenterologist needs to following diagnostic steps.

3.1 Step I

Healthcare professionals should consider assessment for IBS if the person reports having any of the following symptoms for at least 6 months:

- Abdominal pain
- Bloating
- Change in bowel habit.

Bloating means fullness or swelling in the abdomen that often occurs after meals.

During the medical interview, the occurrence of diarrhea, constipation or both problems should be considered. It is very important to ask the patient how many times per day he or she has bowel movements related with visit in the bathroom and about their stool consistency. Sometimes doctors can use the Bristol Scale of stool [2] (Fig. 3.1).

3.2 Step II

All the patients presenting with possible IBS symptoms should be asked if they have any of the following "red flag" indicators and should be referred to secondary care for further investigation if any are present.

- Unintentional and unexplained weight loss
- Rectal bleeding
- A family history of bowel or ovarian cancer
- A change in bowel habit to looser and/or more frequent stools persisting for more than 6 weeks in a person aged over 60 years.

'Red flag' symptoms can be connected to cancer and they are considered as alarming signs. Each time the patient reports even only one of the above symptoms, a specialist's diagnosis is necessary. Special attention must be applied to patients with a positive family history of cancer.

Bristol Stool Chart

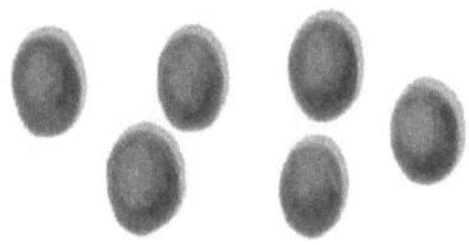	Type 1	Separate hard lumps, like nuts (hard to pass)	Very constipated
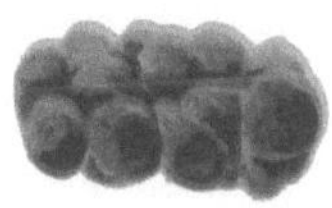	Type 2	Lumpy and sausage like	Slightly constipated
	Type 3	A sasuage shape with cracks in the surface	Normal
	Type 4	Like a sausage or snake, smooth and soft	Normal
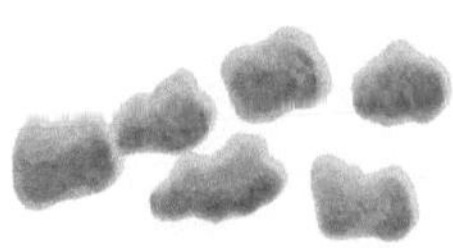	Type 5	Soft blobs with clear-cut edges	Lacking fibre
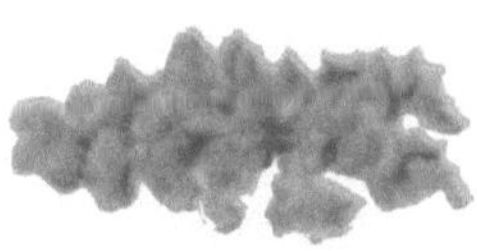	Type 6	Fluffy pieces with ragged edges, a mushy stool	Inflammation
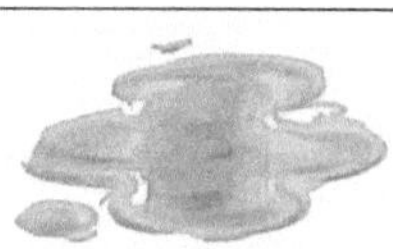	Type 7	Liquid consistency with no solid pieces	Inflammation

Fig. 3.1 Bristol stool chart

3.3 Step III

All the patients presenting with possible IBS symptoms should be assessed and clinically examined for the following "red flag" indicators and should be referred to secondary care for further investigation if any are present.

- Anaemia
- Abdominal masses
- Rectal masses
- Inflammatory markers for inflammatory bowel disease.

In any of these cases, physical examination and imaging studies such as ultrasound should be performed.

If there is a significant concern that symptoms may suggest ovarian cancer, gynecologist's consultation and pelvic examination should also be considered.

3.4 Step IV

The final diagnosis of IBS should be considered only if the person has abdominal pain that is either relieved by defecation or associated with altered bowel frequency or stool form. This should be accompanied by at least two of the following four symptoms:

- Altered stool passage (straining, urgency, incomplete evacuation)
- Abdominal bloating (more common in women than men), distension, tension or hardness
- Symptoms made worse by eating
- Passage of mucus.

Other features such as lethargy, nausea, backache and bladder symptoms are common in people with IBS and may be used to support the diagnosis.

3.5 Step V

3.5.1 Basic Diagnostic Tests

In people who meet the IBS diagnostic criteria, the following fundamental test should be undertaken to exclude other diagnoses and diagnose the IBS:

- Full blood count (FBC)
- Erythrocyte sedimentation rate (ESR) or plasma viscosity

- C-reactive protein (CRP)
- Antibody testing for coeliac disease (endomysial antibodies [EMA] or tissue transglutaminase [TTG]).

To perform above test the venous blood from the peripheral vessels should be drawn.

3.6 Additional Testing

Additional diagnostic testing in patients who meet the IBS diagnostic criteria the tests for celiac disease may be warranted in patients presenting with diarrhea as their predominant symptom. However, the extensive diagnostic testing is unnecessary for patients without alarm symptoms. In some cases addressing disease-related concerns, discussing reasonable treatment goals and expectations, educating and empowering patients, and addressing somatization issues with patients may provide greater benefit than extensive testing. However, even if the above mentioned diagnostic scheme is performed, many of IBS patients need additional clinical tests to exclude the other infections, inflammatory or neoplastic diseases. Additional tests include:

- Abdominal ultrasound
- Rigid/flexible sigmoidoscopy
- Colonoscopy; barium enema
- Thyroid function test
- Faecal ova and parasite test
- Faecal occult blood
- Hydrogen breath test (for lactose intolerance and bacterial overgrowth).

After all of the above steps are completed, IBS may be recognized. However, to systematize the diagnosis of IBS, the Rome III Diagnostic Criteria were introduced in 2006 and Rome IV criteria in 2016 [3].

There were systematic approaches that attempted to classify the then hazy area of FGIDs as early as 1962, when Chaudhary and Truelove published a retrospective review of IBS patients at Oxford, England. Later on, the "Manning Criteria" for IBS were derived from a paper published in 1978 by Manning and colleagues. This seminal classification started a new era and from then on, scientific work on functional gastrointestinal disorders proceeded with increased enthusiasm.

The "Rome process" is an international effort to create scientific data to help in the diagnosis and treatment of functional gastrointestinal disorders (FGIDs), such as IBS, functional dyspepsia and rumination syndrome. The Rome Diagnostic Criteria are set forth by the Rome Foundation, a non- profit organization, under the professional management of Hilliard Associates based in Raleigh, North Carolina.

The Rome criteria have been evolving from the first set, issued in 1989 through the Rome Classification System for FGIDs (1990), the Rome I Criteria for IBS (1992) and the FGIDs (1994), the Rome II Criteria for IBS (1999) and the FGIDs (1999), the Rome III Criteria (2006) and to the most recent Rome IV criteria (2016). Currently the Rome III Diagnostic Criteria for FGIDs is still the 'Gold Standard' for the diagnosis of IBS [4].

3.7 According to Rome III Criteria for IBS Is as Follows

Recurrent abdominal pain at least 3 days/month* in the last 3 months associated with two or more of the following:

- Improvement with defecation
- Onset associated with a change in frequency of stool
- Onset associated with a change in form (appearance) of stool.

Diagnostic Criterion fulfilled for the last 3 months with symptom onset at least 6 months prior to diagnosis.

* Criterion fulfilled for the last months with symptom onset at least months prior to diagnosis.

In pathophysiology research and clinical trials, a pain/discomfort frequency of at least 2 days a week during screening evaluation is recommended for subject eligibility [4].

Patients with IBS are divided into subgroups based on their predominant symptoms:

(a) diarrhea predominant (IBS-D),
(b) constipation predominant (IBS-C),
(c) mixed type with diarrhea and constipation (IBS-M),
(d) undetermined IBS (IBS-U).

Around 75 % of patients are alternators, which illustrates the instability of symptoms over time in the same patient [5].

The Rome III diagnostic questionnaire for IBS contains 10 items and answers to questions are on an ordinal scale with individual frequency thresholds for each question. The qualification of patient to an appropriate subgroup is performed on the answers to questions regarding bowel movements habits, frequency and consistency of stools (Table 3.1).

Table 3.1 ROME III criteria—Questionnaire (Rome Foundation)

1. In the last 3 months, how often did you have discomfort or pain anywhere in your abdomen?	0. Never → 1. Less than one day a month 2. One day a month 3. Two to three days a month 4. One day a week 5. More than one day a week 6. Every day	Skip remaining questions
2. For women: Did this discomfort or pain occur only during your menstrual bleeding and not at other times?	0. No 1. Yes 2. Does not apply because I have had the change in life (menopause) or I am a male	
3. Have you had this discomfort or pain 6 months or longer?	0. No 1. Yes	
4. How often did this discomfort or pain get better or stop after you had a bowel movement?	0. Never or rarely 1. Sometimes 2. Often 3. Most of the time 4. Always	
5. When this discomfort or pain started, did you have more frequent bowel movements?	0. Never or rarely 1. Sometimes 2. Often 3. Most of the time 4. Always	
6. When this discomfort or pain started, did you have less frequent bowel movements?	0. Never or rarely 1. Sometimes 2. Often 3. Most of the time 4. Always	
7. When this discomfort or pain started, were your stools (bowel movements) looser?	0. Never or rarely 1. Sometimes 2. Often 3. Most of the time 4. Always	
8. When this discomfort or pain started, how often did you have harder stools?	0. Never or rarely 1. Sometimes 2. Often 3. Most of the time 4. Always	
9. In the last 3 months, how often did you have hard or lumpy stools?	0. Never or rarely 1. Sometimes 2. Often 3. Most of the time 4. Always	Alternative scale: 0. Never or rarely 1. About 25 % of the time 2. About 50 % of the time 3. About 75 % of the time 4. Always, 100 % time

(continued)

Table 3.1 (continued)

10. In the last 3 months, how often did you have loose, mushy or watery stools?	0. Never or rarely 1. Sometimes 2. Often 3. Most of the time 4. Always	Alternative scale: 0. Never or rarely 1. About 25 % of the time 2. About 50 % of the time 3. About 75 % of the time 4. Always, 100 % time

Criteria for IBS-C (question 9 > 0) and (question 10 = 0)
Criteria for IBS-D (question 9 = 0) and (question 10 > 0)
Criteria for IBS-M (question 9 > 0) and (question 10 > 0)
Criteria for IBS-U (question 9 = 0) and (question 10 = 0)

3.8 Biomarkers for IBS

In 2001, Biomarkers Definitions Working Group defined the term "biomarker" as "a characteristic that is measured and evaluated as an indicator of normal biological processes, pathogenetic processes or pharmacologic responses to a therapeutic agent" [6, 7]. Noninvasive biomarkers are particularly desired, as their application would reduce costs and minimize unnecessary diagnostic tests.

There are several obstacles in the search for relevant biological biomarkers in IBS. They include:

- eterogeneity of symptoms between patients and temporal instability of the symptoms in the same patient
- overlapping of IBS symptoms with other functional gastrointestinal disorders (FGIDs) and more serious organic diseases
- unclear understanding of the pathophysiology of IBS and other disorders [8].

Nevertheless, several new markers in IBS have already been proposed.

1. C-reactive protein (CRP)

C-reactive protein (CRP), a member of pentraxin family is an annular (ring-shaped), pentameric protein found in blood plasma, whose levels rise in response to inflammation. It is an acute-phase protein of hepatic origin that increases following interleukin-6 secretion by macrophages and T cells. Its physiological role is to bind to lysophosphatidylcholine expressed on the surface of dead or dying cells (and some types of bacteria) in order to activate the complement system via the C1Q complex [9].

CRP is synthesized by the liver in response to factors released by macrophages and fat cells (adipocytes) [10–12]. Importantly, it is not related to C-peptide (insulin) or protein C (blood coagulation).

CRP is usually used to assess the degree of inflammation and therapeutic success in diseases such as Crohn's disease and ulcerative colitis. In the study conducted by Hod et al. researchers tried to confirm the hypothesis of elevated high sensitivity CRP (hs-CRP) as a marker of microinflammation in IBS [13]. Hs-CRP levels were higher in IBS patients than HC, but still in the normal laboratory range. This may reflect the low-grade gut inflammation believed to occur in IBS and support its existence. In the study, hs-CRP levels were highest in patients with diarrhea-predominant IBS and in patients with greater disease severity. A cut-off value of 1.08 mg L(-1) demonstrated a sensitivity of 60.2 % and a specificity of 68 % for differentiating IBS from HC. The clinical relevance of CRP values in assessing IBS disease severity or therapy follow-up has not yet been proven.

2. Erythrocyte sedimentation rate (ESR)

The erythrocyte sedimentation rate (ESR), also called a sedimentation rate or Westergren ESR, is the rate at which red blood cells sediment in a period of one hour. It is a common hematology test, and is a non-specific measure of inflammation. The ESR is governed by the balance between pro-sedimentation factors, mainly fibrinogen, and those factors resisting sedimentation, namely the negative charge of the erythrocytes. When an inflammatory process is present, the high proportion of fibrinogen in the blood causes red blood cells to stick to each other [14].

ESR, like CRP, is also hypothesized to be a nonspecific marker for microinflammation [15]. Hauser et al. hypothesized that mild inflammation in IBS patients could be detected by ESR, which could be a sensitive, yet cheap and ubiquitous test [15]. Furthermore, Hauser et al. assumed that ESR would be related with the disease severity index and decreased general and disease-specific health-related quality of life (HRQoL). The preliminary results of a pilot study showed that IBS patients with higher ESR expressed lower disease-specific HRQoL (e.g. they expressed more bowel symptoms, social and emotional disturbances related to disease). No significant correlations were found between ESR and the disease severity as well as general HRQoL.

3. Cortisol

Cortisol belongs to the glucocorticoid class of steroid hormones and is produced in humans by the zona fasciculata of the adrenal cortex within the adrenal gland [16]. It is released in response to stress and low blood-glucose concentration.

Cortisol functions to increase blood sugar through gluconeogenesis, to suppress the immune system, decreases bone formation, and to aid in the metabolism of fat, protein, and carbohydrates [17, 18].

Cortisol is known as the "stress hormone", involved in the body response to stress. The level of cortisol in the blood depends on hypothalamic-pituitary-adrenal (HPA) axis activity. One of the theories suggested that disturbances in HPA axis underlie the development of IBS. Recent studies illustrated that risk factors such as early life trauma and chronic stress increased susceptibility to IBS, with symptoms manifesting after exposure to triggers like changes in enteric flora composition, infection and dietary factors. Therefore, the idea of measuring cortisol levels in these patients and searching for disturbances in the HPA axis seems a logical way of proceeding with research into the origin of disorders [19, 20]. In the study conducted by Kennedy et al., salivary cortisol levels were measured in response to the Trier Social Stress Test (TSST). The authors found greater total cortisol output in response to acute stress in IBS patients compared with healthy subjects [21]. Patients with IBS exhibit sustained HPA axis activity, and often developed different gastrointestinal symptoms in response to acute experimental psychosocial stress.

In 2009, a similar study was performed by FitzGerald et al. The authors measured cortisol levels in women with IBS-D after lumbar puncture as representative of a physical stressor. Results of this study showed an attenuated response of the HPA axis in patients with IBS compared with healthy controls. The impaired tone of the HPA axis was attributed to adaptive changes in brain response to chronic stress to which IBS patients are considered to be more often exposed in comparison with healthy individuals [22]. Women with IBS display blunted adrenocorticotropic hormone and cortisol responses to the lumbar puncture along with a profile of affective responsiveness suggestive of chronic psychosocial stress, although no CRF(CSF) differences between groups were observed.

4. Chromogranin A (CgA)

Chromogranin A (CgA) is a precursor to several functional peptides which negatively modulate the neuroendocrine function of the releasing cell (autocrine signaling) or nearby cells (paracrine signaling). CgA induces and promotes generation of secretory granules including those containing insulin in pancreatic islet beta cells [23]. It is used as an indicator for pancreas and prostate cancer and in carcinoid syndrome [24, 25]. It may also play a role in early neoplastic progression. CgA is cleaved by an endogenous prohormone convertase to produce several peptide fragments.

The chromogranin family recently have been highlighted in the search for the ideal biomarker for IBS. Popularity of this proteins increased since it was discovered that chromogranin family can modulate intestinal inflammation and present active communication between the neuroendocrine and immune system [26]. In the study conducted by Sidhu et al. an elevated CgA serum level in a subset of IBS-D patients was found [27]. The results confirmed the hypothesis about enterochromaffin cell hyperplasia in post-infectious (PI)-IBS patients [28, 29].

The role of chromogranin as an inflammation marker has yet to be proven. In contrast, El-Salhy et al. found no increase in CgA blood level compared with healthy controls and considered that changes in CgA levels in blood are clinically insignificant. Instead, they found reduced density of CgA-containing cells in the duodenum and colon of both IBS-D and IBS-C patients [30]. Because of this finding altered density of intestinal CgA cells as a potential histopathological marker for IBS was proposed [30].

To conclude, a recent study performed by Öhman et al. showed elevated levels of CgA and secretogranins II and III in patients with IBS-D and IBS alternators (IBS-A) [31]. One of the most important observations in this study was that there is a strong negative correlation between the colonic transit time and fecal levels of mentioned granins. This discovery opens the door to new questions and hypotheses regarding the role of fecal granins in IBS.

5. Fecal calprotectin (FC)

Fecal calprotectin (FC) is a biochemical measurement of calprotectin in the stool. Elevated fecal calprotectin indicates the migration of neutrophils to the intestinal mucosa, which occurs during intestinal inflammation, including inflammation caused by inflammatory bowel disease. Under a specific clinical scenario, the test may eliminate the need for invasive colonoscopy or radio-labelled leukocyte scanning [31, 32].

The main diseases that cause an increased excretion of fecal calprotectin are infectious colitis, Crohn's disease, ulcerative colitis, and neoplasms (cancer) [33]. Moreover, the levels of fecal calprotectin seem to be in a normal range in patients with IBS [34].

However, newer studies keep trying to find relevant cutoff FC stool levels that could—with great certainty—distinguish IBS from IBD and reduce unnecessary invasive diagnostic tools. In 2002, Tibble et al. established that the cut-off FC levels of 30 mg/kg combined with Rome I criteria can serve as a clear proof of IBS with no need for further examination [35]. The report published in 2013 showed that FC is confirmed as a highly specific and sensitive biomarker for IBD and the value of 50 mcg/g showed 93 % sensitivity and 94 % specificity in differentiating IBD from IBS [36]. Waugh et al. concluded that FC can be a highly sensitive way of detecting IBD, although there are inevitably trade-offs between sensitivity and specificity, with some false positives (IBS with positive calprotectin) if a low calprotectin cut-off is used. In most cases, a negative calprotectin rules out IBD, thereby sparing most people with IBS from having to have invasive investigations, such as colonoscopy.

In 2014 Chang et al. reported an interesting finding that higher FC levels in IBS patients correlate with disease activity more significantly than serum CRP levels [37]. Findings of elevated FC should be investigated further, because these may increase the sensitivity and specificity of tests performed in the diagnostic algorithm to confirm IBS. Another positive remark on FC is the opinion that FC level correlates with a reduced physical component of health related quality of life (HRQoL) [38]. This means that FC can be used to monitor the response to therapy. Measurement of FC levels should be included in the IBS diagnostic algorithm, regardless of whether it is used to confirm microinflammation and to choose an adequate therapy approach for these patients or to exclude the diagnosis of IBD and minimize unnecessary invasive procedures.

6. Human β-defensin-2 (HBD-2)

Defensins form a family of microbicidal and cytotoxic peptides made by neutrophils. Members of the defensin family are highly similar in protein sequence [39].

HBD-2 is produced by a number of epithelial cells and exhibits potent antimicrobial activity against Gram-negative bacteria and *Candida*, but not Gram-positive *S. aureus*. It has been speculated that HBD-2 may contribute to the infrequency of Gram-negative infections of the skin and lung tissue [40].

In 2009, Langhorst et al. found significantly higher levels of HBD-2 in patients with IBS compared with healthy controls [41]. The results indicate significantly elevated levels of HBD-2 in patients with IBS compared with controls and similar to those with active UC. The results confirm the theory of an activation of the mucosal innate defense system toward a proinflammatory response in IBS patients in the absence of macroscopic signs of inflammation. Langhorst et al. suggested that HBD-2 presents another potential biomarker whose clinical role in IBS has not been adequately investigated so far.

3.9 Conclusion

IBS is still a symptom-based diagnosis disorder that reduces patients' quality of life and which imposes a significant economic burden to the healthcare system. Many healthcare providers view IBS as a static disorder that is hard to define, difficult to diagnose and impossible to treat. These popular views are just several of the most common misconceptions related to the diagnosis and treatment of IBS. The truth, however, is that IBS is a dynamic field characterized by significant changes in diagnostic strategies and therapeutic options over the last decade. The search for a new, cheap and reliable biomarker seems to be the future in diagnosis of IBS.

References

1. National Collaborating Centre for Nursing and Supportive Care (2008) irritable bowel syndrome in adults: diagnosis and management of irritable bowel syndrome in primary care
2. Amarenco G (2014) Bristol Stool Chart: étude prospective et monocentrique de «l'introspection fécale» chez des sujets volontaires [Bristol stool chart: prospective and monocentric study of 'stools introspection' in healthy subjects]. Progrès en Urologie (in French) 24 (11):708–713
3. Rome IV. Functional Gastrointestinal Disorders. Disorders of Gut-Brain Interaction. Douglas A. Drossman, Senior Editor. 2016
4. http://www.romecriteria.org/
5. El-Salhy M (2012) Irritable bowel syndrome: diagnosis and pathogenesis. World J Gastroenterol 18(37):5151–5163

6. Aronson J (2005) Biomarkers and surrogate endpoints. Br J Clin Pharmacol 59(5):491–494. doi:10.1111/j.1365-2125.2005.02435.x.
7. Biomarkers Definitions Working Group (2001) Biomarkers and surrogate endpoints: preferred definitions and conceptual framework. Clin Pharmacol Ther 69(3):89–95. doi:10.1067/mcp.2001.113989
8. Clarke G, Quigley EMM, Cryan JF, Dinan TG (2009) Irritable bowel syndrome: towards biomarker identification. Trends Mol Med 15(10):478–489. doi:10.1016/j.molmed.2009.08.001
9. Thompson D, Pepys MB, Wood SP (1999) The physiological structure of human C-reactive protein and its complex with phosphocholine. Structure 7(2):169–177. doi:10.1016/S0969-2126(99)80023-9
10. Pepys MB, Hirschfield GM (2003) C-reactive protein: a critical update. The Journal of Clinical Investigation 111(12):1805–1812. doi:10.1172/JCI18921
11. Lau DC, Dhillon B, Yan H, Szmitko PE, Verma S (2005) Adipokines: molecular links between obesity and atheroslcerosis. Am J Physiol Heart Circ Physiol 288(5):H2031–H2041. doi:10.1152/ajpheart.01058.2004
12. Mantovani A, Garlanda C, Doni A, Bottazzi B (2008) Pentraxins in innate immunity: from C-reactive protein to the long pentraxin PTX3. J Clin Immunol 28(1):1–13. doi:10.1007/s10875-007-9126-7
13. Hod K, Dickman R, Sperber A et al (2011) Assessment of high-sensitivity CRP as a marker of micro-inflammation in irritable bowel syndrome. Neurogastroenterol Motil 23(12):1105–1110. doi:10.1111/j.1365-2982.2011.01788.x
14. ESR (2013) MedlinePlus: U.S. National Library of Medicine and National Institutes of Health. Retrieved 8 July 2013
15. Hauser G, Tkalcic M, Pletikosic S, Grabar N, Stimac D (2012) Erythrocyte sedimentation rate —possible role in determining the existence of the low grade inflammation in Irritable bowel syndrome patients. Med Hypotheses 78(6):818–820. doi:10.1016/j.mehy.2012.03.020
16. Scott E (2011) Cortisol and stress: how to stay healthy. About.com. Retrieved 29 Nov 2011
17. Hoehn K, Marieb EN (2010) Human anatomy and physiology. Benjamin Cummings, San Francisco. ISBN 0-321-60261-7
18. Chyun YS, Kream BE, Raisz LG (1984) Cortisol decreases bone formation by inhibiting periosteal cell proliferation. Endocrinology 114(2):477–480. doi:10.1210/endo-114-2-477
19. Chitkara DK, van Tilburg MAL, Blois-Martin N, Whitehead WE (2008) Early life risk factors that contribute to irritable bowel syndrome in adults: a systematic review. Am J Gastroenterol 103(3):765–774. doi:10.1111/j.1572-0241.2007.01722.x
20. Blanchard EB, Lackner JM, Jaccard J et al (2008) The role of stress in symptom exacerbation among IBS patients. J Psychosom Res 64(2):119–128. doi:10.1016/j.jpsychores.2007.10.010
21. Kennedy PJ, Cryan JF, Quigley EM, Dinan TG, Clarke G (2014) A sustained hypothalamic-pituitary-adrenal axis response to acute psychosocial stress in irritable bowel syndrome. Psychol Med 44(14):3123–3134. doi:10.1017/s003329171400052x
22. FitzGerald LZ, Kehoe P, Sinha K (2009) Hypothalamic-pituitary-adrenal axis dysregulation in women with irritable bowel syndrome in response to acute physical stress. West J Nurs Res 31(7):818–836. doi:10.1177/0193945909339320
23. Helman LJ, Ahn TG, Levine MA, Allison A, Cohen PS, Cooper MJ, Cohn DV, Israel MA (1988) Molecular cloning and primary structure of human chromogranin A (secretory protein I) cDNA. J Biol Chem 263(23):11559–11563
24. Wu JT, Erickson AJ, Tsao KC, Wu TL, Sun CF (2000) Elevated serum chromogranin A is detectable in patients with carcinomas at advanced disease stages. Ann Clin Lab Sci 30(2):175–178
25. Nikou GC, Lygidakis NJ, Toubanakis C, Pavlatos S, Tseleni-Balafouta S, Giannatou E, Mallas E, Safioleas M (2005) Current diagnosis and treatment of gastrointestinal carcinoids in a series of 101 patients: the significance of serum chromogranin-A, somatostatin receptor scintigraphy and somatostatin analogues. Hepatogastroenterology 52(63):731–741

26. Zhang D, Shooshtarizadeh P, Laventie B-J et al (2009) Two chromogranin a-derived peptides induce calcium entry in human neutrophils by calmodulin-regulated calcium independent phospholipase A2. PLoS One 4(2). doi:10.1371/journal.pone.0004501.e4501
27. Sidhu R, McAlindon ME, Leeds JS, Skilling J, Sanders DS (2009) The role of serum chromogranin A in diarrhoea predominant irritable bowel syndrome. J Gastrointest Liver Dis 18(1):23–26
28. Dunlop SP, Jenkins D, Neal KR, Spiller RC (2003) Relative importance of enterochromaffin cell hyperplasia, anxiety, and depression in postinfectious IBS. Gastroenterology 125 (6):1651–1659. doi:10.1053/j.gastro.2003.09.028
29. Sidhu R, Drew K, McAlindon ME, Lobo AJ, Sanders DS (2010) Elevated serum chromogranin A in irritable bowel syndrome (IBS) and inflammatory bowel disease (IBD): a shared model for pathogenesis? Inflamm Bowel Dis 16(3):361. doi:10.1002/ibd.20982
30. El-Salhy M, Lomholt-Beck B, Hausken T (2010) Chromogranin a as a possible tool in the diagnosis of irritable bowel syndrome. Scand J Gastroenterol 45(12):1435–1439. doi:10.3109/00365521.2010.503965
31. Öhman L, Stridsberg M, Isaksson S, Jerlstad P, Simrén M (2012) Altered levels of fecal chromogranins and secretogranins in IBS: relevance for pathophysiology and symptoms? Am J Gastroenterol 107(3):440–447. doi:10.1038/ajg.2011.458
32. Brophy MB, Nolan EM (2015) Manganese and microbial pathogenesis: sequestration by the mammalian immune system and utilization by microorganisms. ACS Chem Biol 10(3):641–651. doi:10.1021/cb500792b
33. Chen CC, Huang JL, Chang CJ, Kong MS (2012) Fecal calprotectin as a correlative marker in clinical severity of infectious diarrhea and usefulness in evaluating bacterial or viral pathogens in children. J Pediatr Gastroenterol Nutr 55(5):541–547
34. Costa F, Mumolo MG, Ceccarelli L, Bellini M, Romano MR, Sterpi C, Ricchiuti A, Marchi S, Bottai M (2005) Calprotectin is a stronger predictive marker of relapse in ulcerative colitis than in Crohn's disease. Gut 54(3):364–368
35. Tibble JA, Sigthorsson G, Foster R, Forgacs I, Bjarnason I (2002) Use of surrogate markers of inflammation and Rome criteria to distinguish organic from nonorganic intestinal disease. Gastroenterology 123(2):450–460. doi:10.1053/gast.2002.34755
36. Waugh N, Cummins E, Royle P et al (2013) Faecal calprotectin testing for differentiating amongst inflammatory and non-inflammatory bowel diseases: systematic review and economic evaluation. Health Technol Assess 17(55). doi:10.3310/hta17550
37. Chang M-H, Chou J-W, Chen S-M et al (2014) Faecal calprotectin as a novel biomarker for differentiating between inflammatory bowel disease and irritable bowel syndrome. Molecular Medicine Reports. 10(1):522–526. doi:10.3892/mmr.2014.2180
38. Tkalcic M, Pletikosic S, Hausr G (2014) Biological and psychological determinants of health related quality of life in irritable bowel syndrome patients. J Psychosom Res 76(6):516. doi:10.1016/j.jpsychores.2014.03.086
39. Sawai MV, Jia HP, Liu L, Aseyev V, Wiencek JM, McCray PB Jr, Ganz T, Kearney WR, Tack BF (2001) The NMR structure of human beta-defensin-2 reveals a novel alpha-helical segment. Biochemistry 40(13):3810–3816
40. Schröder JM, Harder J (1999) Human beta-defensin-2. Int J Biochem Cell Biol 31(6):645–651
41. Langhorst J, Junge A, Rueffer A et al (2009) Elevated human β-defensin-2 levels indicate an activation of the innate immune system in patients with irritable bowel syndrome. Am J Gastroenterol 104(2):404–410

Chapter 4
Pharmacological and Clinical Treatment of Irritable Bowel Syndrome

Maciej Sałaga and Paula Mosińska

Abstract Irritable bowel syndrome (IBS) is a functional gastrointestinal (GI) disorder with an unknown etiology, which is a growing major concern worldwide. Since the pathophysiology of IBS is barely understood there is no specific treatment for this disorder and numerous treatment options aiming at various pharmacological targets located not only in the GI tract, but also in the central nervous system (CNS) are available. In this chapter we provide an overview on drugs that are currently available for IBS therapy with regard to the type of the disease. We discuss their mechanisms of action, evidences for their effectiveness emerging from clinical trials as well as virtues and drawbacks of the most commonly prescribed medications. Furthermore we highlight the practical aspects of the use of certain drugs, such as possible adverse events and contraindications. Moreover we introduce selected complementary and alternative (CAM) medicine methods that have been proven effective in clinical tests.

Keywords Irritable bowel syndrome · Intestinal transit · Visceral pain G protein-coupled receptors · Ion channels · Hypnosis

List of abbreviations

MOR	μ Opioid receptor
DOR	δ Opioid receptor
KOR	κ Opioid receptor
APN	Aminopeptidase N
CNS	Central nervous system
ClC	Chloride ion channels

The original version of this chapter was revised: Author corrections have been updated and the errors have been corrected. The correction to this chapter is available at https://doi.org/10.1007/978-3-319-49016-8_11

M. Sałaga (✉) · P. Mosińska
Faculty of Medicine, Department of Biochemistry,
Medical University of Lodz, Mazowiecka 6/8, 92-215 Lodz, Poland
e-mail: salaga.maciej@gmail.com

J. Fichna (ed.), *Introduction to gastrointestinal diseases Vol. 1*,
https://doi.org/10.1007/978-3-319-49016-8_4

CFTR	Cystic fibrosis transmembrane conductance regulator
CC	Chronic constipation
CIC	Chronic idiopathic constipation
IBS-C	Constipation-predominant irritable bowel syndrome
DPP IV	Dipeptidyl peptidase IV
ECS	Endocannabinoid system
EOS	Endogenous opioid system
GI	Gastrointestinal
GC-C	Guanylate cyclase C
HEK cells	Human embryonic kidney cells
EC cells	Enterochromaffin cells
5-HT	Serotonin
SERT	Serotonin-selective reuptake transporter
SCBM	Spontaneous complete bowel movement
TPH1/2	Tryptophan hydroxylase 1/2
5-HT2B	Type 2B serotonin receptor
5-HT3	Type 3 serotonin receptor
5-HT4	Type 4 serotonin receptor

4.1 Introduction

Irritable bowel syndrome (IBS) is a complex and multifactorial disease with multiple molecular mechanisms involved in its pathophysiology. Consequently, the exact mechanism responsible for the development of IBS remains unrevealed what significantly hinders the search for the new medications. Nevertheless meaningful efforts have been put on the development of drugs that secure fundamental therapeutic goals of the treatment of IBS, such as improvement of unbearable symptoms accompanied with augmentation of patients quality of life. Intensive research in this field lead to the development of several treatment options tailored to the specific groups of patients based on the type of IBS or gender (please see below for more details). Since IBS is not accompanied with any known organic changes in the gut, and none of these treatment options includes invasive procedures, the IBS patients must rely solely on the pharmacotherapy and lifestyle modification. Of note, the present-day prospect of anti-IBS therapies as well as drugs available on the market undergoes constant variations with some compounds being withdrawn and other ones entering it in a relatively short time span.

In this chapter we provide an overview on the pharmacological targets for small-molecule anti-IBS drugs including serotonin receptors, chloride ion channels, guanylate cyclase C (GC-C), endogenous opioid system (EOS), cystic fibrosis transmembrane conductance regulator (CFTR) and somatostatin-2 receptors (Fig. 4.1). We also discuss the virtues and drawbacks of the most commonly

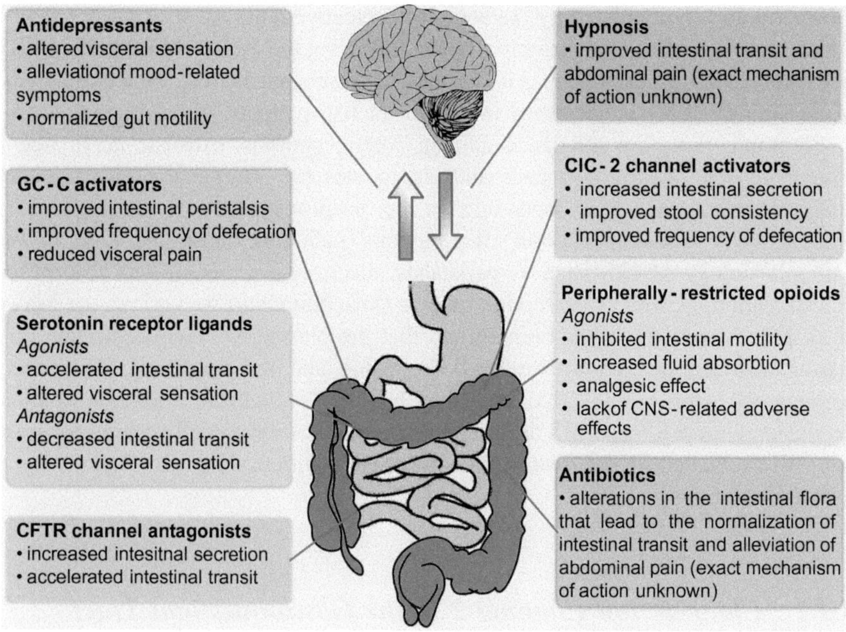

Fig. 4.1 An overview on pharmacological targets for clinically validated anti-IBS drugs/interventions. *ClC-2* type 2 chloride ion channels; *CFTR* cystic fibrosis transmembrane conductance regulator; *CNS* central nervous system; *GC-C* Guanylate cyclase C

prescribed medications. Moreover, we introduce selected complementary and alternative (CAM) medicine methods that have been proven effective in clinical tests.

4.2 Pharmacological Targets

4.2.1 Serotonin System in the Gastrointestinal Tract

Serotonin, which is a derivative of the exogenous amino acid tryptophane is synthesized and stored in the enteric enterochromaffin (EC) cells located in the intestinal mucosa. In the human gastrointestinal (GI) tract EC cells are the most abundant in the duodenum and rectum and the scarcest in the ileum. On the other hand activated mast cells may also contribute to the synthesis and secretion of serotonin. The concentration of serotonin is relatively low in the duodenum and ileum (1.4 and 0.6 nmol/mg protein, respectively) and gradually rises in the colon reaching 45 nmol/mg protein in the rectum [1]. The synthesis of serotonin in the gut requires tryptophan hydroxylase 1 (TPH1), which is a rate limiting enzyme in this process [2]. In neurons serotonin is synthesized by an isoform of tryptophan hydroxylase, TPH2 [3]. Additionally Moreover, the availability of serotonin in the

GI tract is locally regulated by the serotonin-selective reuptake transporter (SERT) which removes it from the interstitial space following the release by EC cells. SERT is expressed by all epithelial cells of the intestinal mucosa [2]. Noteworthy, the expression of SERT is decreased in the gut of IBS patients [4].

Serotonin receptors are distributed on enteric neurons, extrinsic nerve fibers, smooth muscle cells, goblet cells and enterocytes [2]. They can exert excitatory and/or inhibitory activities depending on the receptor type. Serotonin secreted by the EC cells mediates various GI functions including those involved in the pathophysiology of IBS such as peristalsis, electrolyte secretion and absorption, vasodilatation, as well as perception of pain (for comprehensive review, see Mawe et al. [2]). Moreover, it has been shown that the plasma 5-HT concentration correlates with colonic motility under both fasting and fed conditions [1]. Hence, perhaps not surprisingly, serotonin system has been recognized as one of the most promising targets for anti-IBS drugs and stimulation of serotonin receptors has been clinically validated for the treatment of disorders manifested by disturbed intestinal motility and pain.

4.2.2 Chloride Ion Channel 2 in the Gastrointestinal Tract

Chloride ion channels (ClC) constitute an evolutionarily well-conserved family of voltage-gated channels that are structurally unrelated to the other known voltage-gated channels. To date several types of ClC have been identified, including ClC-0, ClC-1, ClC-2, ClC-4 and ClC-5 [5]. ClCs are involved in the regulation of the excitability of neurones, smooth muscle cells, cell volume control and transepithelial salt transport.

ClC-2 is a member of the ClC family that is ubiquitously expressed in mammalian tissues and has been found in both small and large intestinal epithelial cells as well as on GI parietal cells [6]. In the physiological membrane conditions the channel is closed; however, it may be activated by hyperpolarisation, cell swelling as well as acidic extracellular pH [5]. Chloride secretion is responsible for maintenance of mucosal hydration throughout the GI tract, and chloride transport is also pivotal in the regulation of fluid secretion into the intestinal lumen [6]. Activation of ClC-2 enables translocation of chloride ions across the membrane followed by the release of sodium and water into the gut lumen. The influx of fluid into the intestine promotes GI motility and increases the colonic transit together with the number of spontaneous bowel movements. The surplus of water is absorbed by the colonic epithelial tissue what limits the emergence of diarrhea [6]. Taken together, ClC-2 has been validated as a target for the treatment of chronic idiopathic constipation as well as constipation-predominant IBS (IBS-C) [7].

4.2.3 Guanylate Cyclase-C in the Gastrointestinal Tract

GC-C is a transmembrane enzyme, belonging to the protein family synthesizing one of the most common and important secondary messengers—cyclic guanosine monophosphate (cGMP) [8]. There are seven members of the GC family (GC-A–G); however, only GC-C has been validated as a pharmacological target for the treatment of GI pathologies. The endogenous activators of GC-C include peptides, guanylin and uroguanylin which play important function in the maintenance of gut homeostasis. Moreover, GC-C is known as a target protein for heat-stable enterotoxins produced by numerous enteric pathogens that colonize intestines, including Escherichia coli, Citrobacter freundii, Vibrio cholerae and Yersinia enterocolitica [8, 9]. GC-C is expressed on the brush border of intestinal cells along the small and large intestine. Its expression is regulated by intestine-specific transcription factor Cdx2 and is higher in the crypt of the colonic mucosa compared to the crypt of the small intestine [10, 11].

Activation of GC-C leads to the increase of the intracellular level of cGMP, what causes activation of the cGMP-dependent protein kinase II (PKG II). PKG II by phosphorylation of cystic fibrosis transmembrane conductance regulator (CFTR) ion channel induces secretion of chloride and HCO_3-into the intestinal lumen. Moreover, cGMP reduces absorption of Na^+ ions by Na^+/H^+ exchanger [11]. All these events lead to the accumulation of osmotically active molecules in the intestines what causes massive influx of water and increased excretion [12]. The pro-excretory properties of GC-C activators have been exploited in the development of synthetic GC-C agonists that are used in the treatment of functional GI disorders manifested by chronic constipation, such as IBS-C [8, 13, 14].

4.2.4 Cystic Fibrosis Transmembrane Conductance Regulator in the Gastrointestinal Tract

CFTR is a cyclic AMP (cAMP)-regulated ion channel that transfers chloride and thiocyanate ions through the membrane of various types of epithelial cells. It consists of two transmembrane domains linked by the R domain whose phosphorylation by the protein kinase A (PKA) leads to the opening of the gate for the ions [15]. The expression of CFTR alters throughout the GI tract. The lowest level is observed in the mucosal epithelium of the stomach. In the ileum the expression is relatively high and exhibits decreasing gradient along the crypt axis [16]. Furthermore, a small subpopulation of the cells of yet unrevealed function has been shown to express CFTR in the duodenum and jejunum [16]. In the colon the expression of CFTR is the highest in the base of the crypts and resembles the pattern occurring in the small intestine [16]. In the physiological conditions CFTR is responsible for the proper production of the mucus, secretion of fluids into the intestinal lumen and has a strong impact on GI motility and excretion. Knock out of Cftr gene impairs the intestinal transit and lowers the volume of fluids in the gut [17]. On the other hand, CFTR upregulates some of the genes associated with the GI inflammation and stimulates accumulation of mast cells in the intestinal smooth muscle tissue [18]. In line, cystic fibrosis patients (possessing mutation on the Cftr gene) reveal prolonged intestinal transit compared to healthy controls [18].

CFTR is one of the most important factors involved in the proper formation of the intestinal mucus, which constitutes a niche for the growth of intestinal microbiota. Thus, perhaps not surprisingly, loss of CFTR is associated with significant decreases in GI bacterial community richness, evenness and diversity as well as reduced abundance of protective species, including a multitude of Lactobacillales members [19].

The properties and functions of CFTR made it an attractive target for the treatment of disorders accompanied with deregulated motility and abdominal pain.

4.2.5 Endogenous Opioid System in the Gastrointestinal Tract

Endogenous opioid system (EOS) consists of three main types of opioid receptors, namely μ, κ and Δ (MOR, KOR and DOR respectively). Their respective endogenous ligands, endorphins, dynorphins and enkephalins as well as enzymatic machinery dedicated to their degradation, including various proteases [e.g. aminopeptidase N (APN) and dipeptidyl peptidase IV (DPP IV)] [20]. Opioid receptors are widely distributed in the human body. All opioid receptor subtypes have been localized in the gastrointestinal tract of many mammalian organisms. In human body the highest concentration of MOR in the human body has been detected in the myenteric and submucosal plexuses, on immune cells in the lamina propria and ileal longitudinal muscle. DOR was detected in the enteric ganglia and fibers of esophagus, duodenum, ileum, cecum as well as in the proximal, and distal colon. KORs were localized on the myenteric and submucosal neurons, smooth muscle fibres as well as mucosa in rats [21]. Furthermore, opioid receptors were also found in high amounts on lymphocytes and macrophages, which suggest their involvement in the modulation of function of these cells [22]. EOS is crucially involved in numerous physiological processes, including pain signaling in the central and the peripheral nervous system, and respiration.

In the GI tract opioid receptors play a major role in the regulation of GI transit, secretion and immune responses. The major effects of opioid receptor agonists in the GI tract are reduction of intestinal contractility and impairment of peristalsis caused by blockade of neurotransmitter release [22]. Moreover opioids promote water and electrolyte absorption thus decreasing the volume of intestinal content and frequency of excretion. On the other hand, Moreover, both natural and synthetic opioid agonists exhibit potent analgesic effects and decrease abdominal pain in both physiological and pathophysiological conditions [23–25]. To date, several EOS-targeting compounds reached the market and found a place in the clinical treatment of GI-related conditions (e.g. loperamide, alvimopan, oxycodone, racecadotril; for comprehensive review please see Mosinska et al. [21]).

4.3 Pharmacological Treatment of Diarrhea-Predominant IBS (IBS-D)

4.3.1 Alosetron

Alosetron is a 5-HT3 receptor antagonist effective the treatment of IBS-D in women. It is a therapeutic agent with a limited use and is available only for severe and unresponsive to other agents IBS-D cases. It improves pain and discomfort as well as stool frequency and urgency [7, 26]. Alosetron was approved by the US Food and Drug Administration (FDA) in 2000, after a seven month review process. However, eight months later it was removed from the market following reports of serious complications, such as severe constipation and ischemic colitis that, in several cases, lead to a surgery. In 2002 FDA reconsidered the case of alosetron and reintroduced it to the market under a risk management plan with a lower recommended starting dose of 0.5 mg twice daily [7]. In 2005 and 2007 Chang et al. [27] and Krause et al. [28] respectively, have shown the effectiveness of alosetron in the treatment of IBS-D both in men ($n = 662$) and women ($n = 705$) reporting low incidence of serious adverse events. The recent 9-year evaluation of trends in alosetron postmarketing safety under the risk management program indicate that incidence of ischemic colitis and constipation remain rare and stable, at approximately 1 case/1000 patient-years [29]. The indications for alosetron in women with severe IBS-D include: (i) chronic IBS symptoms (generally lasting 6 months or longer), (ii) the absence of anatomic or biochemical abnormalities of the GI tract excluded, (iii) disability or restriction of daily activities due to IBS and (iv) no adequate response to conventional therapy.

4.3.2 Ramosetron

Ramosetron is a potent and selective 5-HT3 receptor antagonist, which has been initially developed for the treatment of nausea and vomiting [30]. Clinical studies showed that ramosetron is effective against IBS-D. In a double-blind, placebo-controlled, parallel-group study of 418 male and female patients with IBS-D ramosetron increased the monthly responder rates of IBS symptoms compared to placebo [31]. In another 12-week randomized controlled trial of 539 patients, a positive response to treatment was reported by 47 % [32]. Furthermore, the drug was active after oral administration. A long-term efficacy for overall improvement of IBS symptoms was also demonstrated. Seven % of patients reported adverse events after ramosetron treatment; however, no serious adverse events (severe constipation, ischemic colitis), were reported for long-term treatment with ramosetron [33]. Ramosetron is only licensed for use in Japan and selected Southeast Asian countries (e.g. India).

4.3.3 Loperamide

Loperamide is a synthetic peripherally-restricted MOR agonist, which does not cross the blood-brain barrier. It decreases gastric emptying, slows peristalsis, delays intestinal transit and relaxes the segmental colonic smooth muscles. On the other hand it increases fluid absorption and inhibits intestinal secretion of electrolytes [34]. In IBS-D loperamide combats diarrhea and reduces stool frequency; however, it has only limited effect on abdominal pain. Clinical features of loperamide are well-established, the drug is safe and effective hence it is often recommended as a first-line therapy for functional GI disorders accompanied with diarrhea in adults. At high doses loperamide may induce constipation; therefore, the treatment starts with a relatively low dose (approx. 2 mg) and then it is titrated up or down based on the symptoms [7, 34]. Clinical studies demonstrated that loperamide is well tolerated in a 5-week therapy [35].

4.3.4 Trimebutine

Trimebutine (used in the form of trimebutine maleate) is a weak agonist of peripheral MOR, KOR and DOR receptors, which also exhibit antimuscarinic properties [36]. Trimebutine accelerates gastric emptying, induces premature phase III of the migrating motor complex in the intestine and modulates the contractile activity of the colon [37]. Clinically, trimebutine has been shown to alleviate both acute and chronic abdominal pain in patients with IBS and it may also be used in children with abdominal pain. Recently, Karabulutu et al. [36] evaluated the effect of trimebutine versus non-medication in 345 children and adolescents demonstrating the effectiveness (94.9 % patients in trimebutine group experienced significant relief) [36]. The indications for trimebutine include: (i) IBS, (ii) abdominal pain and abdominal cramping and (iii) dyspepsia. It may be administered in multiple doses per day with the maximal total daily dose of 600 mg.

4.3.5 *Eluxadoline*

Eluxadoline is a peripherally-restricted mixed MOR agonist and DOR antagonist approved by FDA in May 2015 [34, 38]. In 2013 a phase II clinical trial (n = 807) demonstrated the effectiveness of eluxadoline versus placebo against global IBS-D symptoms [39]. Patients receiving a drug were significantly more likely to meet the US FDA response end point during the full 12 weeks of the study than those receiving placebo. Eluxadoline was well tolerated with a low incidence of constipation. Phase III trials (n = 2428 patients in total) confirmed these results and showed that treatment with eluxadoline (75 or 100 mg twice daily) lead to simultaneous improvement in abdominal pain and stool consistency on the same day for ≥50 % of days over weeks 1–12 and 1–26 of the study (for more details please see Nee et al. [34]). On the other hand a nonsignificant improvement in worst abdominal pain scores in those who received eluxadoline compared to placebo was observed. Common adverse effects in the two phase III clinical trials were nausea, headache, nasopharyngitis, abdominal pain and constipation but rates of discontinuation due to constipation were low (approx. 1.5 %) for both eluxadoline and placebo [40]. Known contraindications to the treatment with eluxadoline include: (i) biliary duct obstruction, or sphincter of Oddi disease or dysfunction, (ii) alcohol abuse or addiction, or patients who drink more than three alcoholic beverages per day, (iii) a history of pancreatitis or structural diseases of the pancreas, including known or suspected pancreatic duct obstruction, (iv) a history of chronic or severe constipation or known or suspected mechanical gastrointestinal obstruction.

4.3.6 *Rifaximin*

Rifaximin is another drug for IBS-D approved by FDA in 2015. It is a nonabsorbable, semisynthetic antibiotic belonging to the rifamycin family. The use of antibiotics for the treatment of IBS emerged from the observation that gut microflora differs between IBS and general population. Furthermore epidemiological data reveal that up to 31 % of IBS cases are caused by an episode of gastroenteritis [34]. Rifaximin targets the β-subunit of bacterial RNA polymerase which is responsible for the transcription process [41]. It does not affect the overall composition of the microbiota but appear to influence mainly potentially detrimental species such as *Clostridium* sp. and increases the presence of some species, such as Faecalibacterium prausnitzii [42].

Clinical trials suggest that the drug can reduce global IBS symptoms, improve bloating, abdominal pain, and stool consistency in patients with non-constipated IBS [43]. While other anti-IBS therapies require daily administration to maintain their efficacy, 2-week rifaximin treatment can achieve symptom improvement that persists up to 12 weeks post-treatment [44]. However, in the clinical trial it has been shown that 64 % of patients who underwent a 2-week therapy with rifaximin (550 mg) develop a relapse in the 18 weeks follow up hence FDA recommends a 14-day therapy with rifaximin at the dose of 550 mg (orally), three times a day. In case of the recurrence of the symptoms therapy may be repeated for another 14 days. Rifaximin is well-tolerated both after single and repeated treatments with a side effect profile comparable to that of placebo. The most common adverse events caused by rifaximin are headache, upper respiratory infection, nausea, nasopharyngitis, diarrhea and abdominal pain.

4.3.7 Crofelemer

Crofelemer is a plant-derived drug originating from Croton lechleri, which belongs to the proanthocyanidin family. It was approved by FDA for the treatment of diarrhea associated with anti-HIV drugs [45]. It simultaneously targets two distinct channels, CFTR and calcium-activated chloride channel, both responsible for chloride and fluid secretion in the GI tract. Although it has been shown that crofelemer did not produce significant improvement in stool consistency, stool frequency, urgency and adequate relief it increased the number of pain-free days in female IBS-D patients after 1 and 3 month therapy and was well tolerated [46]. Further studies evaluating the analgesic potential action of this drug are needed to draw a clear conclusion on its therapeutic potential.

4.3.8 Antidepressants

Antidepressants are commonly used in IBS-D. There are many Plethora of evidences for point to the link between mood-related disorders and functional GI diseases. Emotional fluctuations that often occur in distressed patients correlate with IBS symptoms. Moreover, IBS patients are more likely to develop psychiatric disorders (depression, anxiety) and dementia [47, 48].

The bidirectional communication between the brain and the gut, so called brain-gut axis, may be exploited therapeutically in IBS patients. Some of the tricyclic antidepressants, selective serotonin re-uptake inhibitors and serotonin-norepinephrine reuptake inhibitors have already been employed in the treatment of IBS and proved effective in symptom relief via mood stabilization, modulation of pain perception and amelioration of GI motility and secretion. A recent meta-analysis confirmed the efficacy of antidepressants, including tricyclic antidepressants, in the treatment of IBS symptoms [49]. In a randomized, double-blind, placebo controlled study low dose amitriptyline (10 mg) successfully ameliorated IBS-D symptoms [50]. Fifty out of 54 patients completed an intention-to-treat study; 68 % of those receiving amitriptyline had a complete response defined as a loss of all symptoms over a 2 month trial period compared to only 28 % of the controls. Adverse effects were similar between the two groups.

4.4 Pharmacological Treatment of Constipation-Predominant Irritable Bowel Syndrome (IBS-C)

4.4.1 Polyethylene Glycol (PEG) 3350

The first-line therapy for patients suffering from IBS-C involves laxatives and dietary fibers. Although this approach may effectively and safely combat slowed intestinal transit and constipation, it does not alleviate pain symptoms [51]. The effect of PEG 3350 plus electrolytes (PEG+E) on IBS-C has been tested in a randomized, double-blind, placebo controlled study by Chapman et al. [52]. One hundred thirty four patients received the treatment or placebo for 28 days. PGE+E was superior than placebo as assessed by spontaneous bowel movements (the primary endpoint), responder rates, stool consistency, and straining. There was no difference between PGE+E versus placebo in the mean severity score for abdominal discomfort/pain. PEG+E constitutes a well tolerable and effective treatment that should be considered suitable for use as a first-line treatment in functional GI disorders manifested by chronic constipation.

4.4.2 Tegaserod

Tegaserod is a partial 5-HT4 receptor agonist that reduces visceral sensitivity and stimulates the secretion of chloride from epithelial cells. It has been approved by FDA in 2002 and subsequently withdrawn from the market in 2007 due to possible adverse cardiovascular effects (heart attack and stroke) [26]. The putative adverse events caused by tegaserod most likely result from its non-selective binding to other serotonin receptors, such as 5-HT1, 5-HT2a and 5-HT2b [53]. FDA had been criticized for this decision and ultimately reconsidered it and allowed for reintroduction of tegaserod under an investigational new drug protocol for IBS-C and chronic idiopathic constipation in women younger than 55 who are not at risk for certain cardiovascular events [53, 54]. The effect of tegaserod on IBS-C in women has been evaluated in a large ($n = 661$) randomized, controlled trial [55]. It provided significant improvement and satisfactory relief of IBS symptoms over 4 weeks of treatment in 43.3 % of IBS-C patients. The most frequent adverse events leading to study discontinuation in tegaserod-treated patients were diarrhea (1.5 %) and abdominal pain (0.9 %). Although long-term safety of tegaserod was investigated in a prospective study suggesting that treatment was safe over a 12-month period tegaserod was not approved for use in the EU due to the opinion that its benefits does not outweigh its risks [56].

4.4.3 Prucalopride

Prucalopride, which belongs to benzofurans, is a selective agonist of 5-HT4 receptor that exhibits prokinetic effect in the GI tract. It stimulates colonic peristalsis, which provides the main propulsive force for defecation. On the contrary to other 5-HT4, it does not induce cardiovascular adverse events, which may be attributed to its high selectivity over other types of 5-HT receptors and ion channels. Clinical trials with prucalopride (1974 patients in total; both men and women) demonstrated a significant increase in the proportion of patients achieving at least three spontaneous complete bowel movements (SCBMs) per week compared with placebo [57–59]. Response rates ranged from 24 to 28 % with 4 mg prucalopride, and 9.6–12 % with placebo. Clinically relevant improvement was also demonstrated in other measures, including satisfaction with bowel function, perception of the severity of constipation as well as quality of life. It should be also underlined that prucalopride is not effective in children with functional constipation, as showed by Mugie et al. [60]. Regardless of the patient's age, prucalopride is well tolerated with no impact on the cardiovascular system [26]. The most frequently reported adverse events include headache, abdominal pain, nausea and diarrhea. Prucalopride has been approved in Europe for both men and women; however, it has not been allowed for sale in the USA.

4.4.4 Linaclotide

Linaclotide is a 14 amino acid peptide agonist of GC-C which has been approved by FDA for the treatment of IBS-C in 2012 and to date is considered as a first-in-class drug by majority of gastroenterologists. It is characterized by low bioavailability (approx. 0.1 %), what enables local action in the intestines. Linaclotide activates GC-C and causes an increase in the level of intracellular cGMP with concomitant upregulation of HCO_3-and chloride ions what results in an increased secretion and acceleration of intestinal transit [8]. Clinical data demonstrated that linaclotide improves severity of abdominal pain as well as bowel movements in IBS-C patients (for more details please see Jarmuz et al. [8]). Phase I trial showed that linaclotide provides relief and is well tolerated in 42 patients [61]. Rao et al. [62] reported the effects of 12-week treatment with linaclotide in IBS-C patients ($n = 800$). One-third of patients receiving linaclotide reached the

FDA-recommended primary endpoint (improvement of ≥30 % from baseline in the average of the daily worst abdominal pain score on a standardized scale and an increase of at least 1 CSBM from baseline in the same week for at least 6 of first 12 weeks of treatment). During the withdrawal period patients receiving linaclotide experienced sustained decrease of abdominal pain while placebo-treated patients had a gradual increase of the pain score. In another clinical study linaclotide administered orally improved global IBS-C symptoms during 26-week therapy [63]. In line with the previous studies linaclotide induced significant relief in approx. one-third of the patients. Abdominal discomfort, fullness, cramping and bloating were also significantly improved. The most common adverse effect, which leads to discontinuation of the medication with linaclotide is diarrhea, occurring in approximately 5 % of patients [64].

4.4.5 Lubiprostone

Lubiprostone (approved by FDA in 2008 to treat IBS-C) is a bicyclic fatty acid derived from prostaglandin E1 that activates ClC-2 chloride channels located on the apical area of GI epithelial cells. It is poorly absorbed from the gut what facilitates its local activity in the GI tract [65]. Although it is widely accepted that lubiprostone acts via apical ClC-2 channels, recently some novel insights into its mechanism of action have been demonstrated. It was shown that lubiprostone, not only activates apical ClC-2 channels but also induces the internalization of basolateral ClC-2 into the cytoplasm with concomitant trafficking of CFTR and chloride/hydrogen carbonate exchanger PAT-1 to the apical membrane [66]. At the molecular level events triggered by lubiprostone leads to the increased luminal secretion of chloride and decreased absorption of this ion by basolateral ClC 2 channels. These events soften the stool, increase motility, and promote SCBMs.

In clinical trials lubiprostone was shown to improve SCBMs frequency after 1 week of therapy. Of note, some of the patients (approx. 55 %) experienced a relief in the first day of the treatment. Improved stool consistency, straining, and constipation severity, as well as patient-reported assessments of treatment effectiveness, were also reported [67–69]. The most common adverse events of lubiprostone are nausea, diarrhea, headache and abdominal distention.

The recommended dose of lubiprostone is 8 μg twice daily; however, this dose might be increased if the symptoms do not improve [70]. Of note, some of the patients experience significant relief in all symptoms only after 1 month of the treatment what has to be taken into consideration by specialists planning therapy. Lubiprostone is contraindicated in patients exhibiting chronic diarrhea, bowel obstruction, or IBS-D and is not approved for the use in children.

4.5 Conclusions and Future Perspectives

Significant improvement of patient's quality of life, which is an ultimate goal of all anti-IBS therapies, can be only achieved if the drug/intervention used satisfies several clearly defined conditions. On top of all, a sufficient efficacy and acceptable safety in subjects with IBS is required. Furthermore, selection of a medicine should take into account patients lifestyle, other medicines that are possibly used in the same time and all other contraindications. These issues may particularly affect patient's adherence to medication. To enhance the compliance and satisfaction of the patient, improvement of all symptoms, including diarrhea and/or constipation as well as abdominal pain should be secured by one a single drug. Moreover, reliable estimation of possible drug-related adverse events, such as nausea or headache is critical for proper selection of the drug(s).

Noteworthy, the race for novel anti-IBS medications is always on and patients can be reassured that several novel, superior compounds will enter the market in the next few years (please see Mosinska et al. [71] and Deiana et al. [72] for detailed information on experimental drugs).

As shown in this chapter, a significant number of highly effective and safe synthetic and semi-synthetic drugs is currently available on the market for all types of IBS, often tailored to the needs of particular groups of patients. However, it has to be underlined that there is also a significant group of non-responders who struggle to find an appropriate method of treatment. These people often reach to the complementary and alternative therapies that, as proven clinically, may also provide a long-awaited relief. There are several herbal preparations that may provide at least transient relief, such as peppermint oil capsules (for detailed information please see [73, 74]). Moreover, acupuncture, which is commonly used in China, has emerged as a new potential anti-IBS therapy. However, based on the available data which is often contradictory, it is difficult to state a firm conclusion on its effectiveness and clinical relevance [75–77]. One of the most intriguing forms of therapies is hypnosis, which has been evaluated in large clinical trials (n = 1000) demonstrating its safety and potency in refractory IBS [78]. The mechanism of this method is still unexplained; however, it holds a great promise for many IBS sufferers who do not experience sufficient relief with a standard therapy [79].

Box 4.1 Anti-IBS drugs in the nutshell

Drug	Key information
IBS-D	
Alosetron	• $5\text{-}HT_3$ receptor antagonist effective the treatment of IBS-D in women • Indications include: (i) chronic IBS symptoms (generally lasting 6 months or longer), (ii) the absence of anatomic or biochemical abnormalities of the GI tract, (iii) disability

(continued)

(continued)

Drug	Key information
	or restriction of daily activities due to IBS and (iv) no adequate response to conventional therapy
Ramosetron	• Licensed for use only in Japan and selected Southeast Asian countries (e.g. India)
Loperamide	• At high doses loperamide may induce constipation • Treatment starts with a relatively low dose (approx. 2 mg) and then it is titrated up or down based on the symptoms
Trimebutine	• indications include: (i) IBS, (ii) abdominal pain and abdominal cramping and (iii) dyspepsia • May be administered in multiple doses per day with the maximal total daily dose of 600 mg
Eluxadoline	• Contraindications to the treatment include: (i) biliary duct obstruction, or sphincter of Oddi disease or dysfunction, (ii) alcohol abuse or addiction, or patients who drink more than three alcoholic beverages per day, (iii) a history of pancreatitis or structural diseases of the pancreas, including known or suspected pancreatic duct obstruction, (iv) a history of chronic or severe constipation or known or suspected mechanical gastrointestinal obstruction
Rifaximin	• Recommended is 14-day therapy with the dose of 550 mg (orally), three times a day • Therapy may be repeated for another 14 days • The most common adverse events are headache, upper respiratory infection, nausea, nasopharyngitis, diarrhea and abdominal pain
IBS-C	
Polyethylene glycol 3350 + electrolytes	• Well tolerable and effective treatment that should be considered suitable for use as a first-line treatment in functional GI disorders manifested by chronic constipation
Tegaserod	• Long-term safety of tegaserod was investigated in a prospective study suggesting that treatment was safe over a 12-month period • Not approved for use in the EU due to the opinion that its benefits do not outweigh its risks
Prucalopride	• The most frequently reported adverse events include headache, abdominal pain, nausea and diarrhea • Approved in Europe for both men and women; however, it has not been allowed for sale in the USA
Lubiprostone	• Recommended dose is 8 μg twice daily; however, this dose might be increased if the symptoms do not improve • Contraindicated in patients exhibiting chronic diarrhea, bowel obstruction, or IBS-D and is not approved for the use in children

Acknowledgments The author is supported by the Medical University of Lodz [502-03/1-156-04/502-14-140 to M Salaga] and the National Science Centre [#UMO-2015/16/T/NZ7/00031 and #UMO-2013/11/N/NZ7/02354 to M Salaga]. This study is also sponsored by the Polpharma Scientific Foundation. The author have no other relevant affiliations or financial involvement with any organization or entity with a financial interest in or financial conflict with the subject matter or materials discussed in the manuscript apart from those disclosed.

Disclosures The authors have nothing to disclose.

References

1. Spiller R (2008) Serotonergic agents and the irritable bowel syndrome: what goes wrong? Curr Opin Pharmacol 8:709–714
2. Mawe GM, Hoffman JM (2013) Serotonin signalling in the gut–functions, dysfunctions and therapeutic targets. Nat Rev Gastroenterol Hepatol 10:473–486
3. Margolis KG, Stevanovic K, Li Z, Yang QM et al (2014) Pharmacological reduction of mucosal but not neuronal serotonin opposes inflammation in mouse intestine. Gut 63:928–937
4. Coates MD, Mahoney CR, Linden DR, Sampson JE et al (2004) Molecular defects in mucosal serotonin content and decreased serotonin reuptake transporter in ulcerative colitis and irritable bowel syndrome. Gastroenterology 126:1657–1664
5. Jentsch TJ, Gunther W (1997) Chloride channels: an emerging molecular picture. BioEssays 19:117–126
6. Jiang C, Xu Q, Wen X, Sun H (2015) Current developments in pharmacological therapeutics for chronic constipation. Acta Pharm Sin B 5:300–309
7. Lacy BE, Chey WD, Lembo AJ (2015) New and Emerging treatment options for irritable bowel syndrome. Gastroenterol Hepatol (NY) 11:1–19
8. Jarmuz A, Zielinska M, Storr M, Fichna J (2015) Emerging treatments in neurogastroenterology: perspectives of guanylyl cyclase C agonists use in functional gastrointestinal disorders and inflammatory bowel diseases. Neurogastroenterol Motil 27:1057–1068
9. Maneerattanaporn M, Chang L, Chey WD (2011) Emerging pharmacological therapies for the irritable bowel syndrome. Gastroenterol Clin North Am 40:223–243
10. Birbe R, Palazzo JP, Walters R, Weinberg D et al (2005) Guanylyl cyclase C is a marker of intestinal metaplasia, dysplasia, and adenocarcinoma of the gastrointestinal tract. Hum Pathol 36:170–179
11. Lima AA, Fonteles MC (2014) From Escherichia coli heat-stable enterotoxin to mammalian endogenous guanylin hormones. Braz J Med Biol Res 47:179–191
12. Steinbrecher KA, Cohen MB (2011) Transmembrane guanylate cyclase in intestinal pathophysiology. Curr Opin Gastroenterol 27:139–145
13. Busby RW, Bryant AP, Bartolini WP, Cordero EA et al (2010) Linaclotide, through activation of guanylate cyclase C, acts locally in the gastrointestinal tract to elicit enhanced intestinal secretion and transit. Eur J Pharmacol 649:328–335
14. Busby RW, Kessler MM, Bartolini WP, Bryant AP et al (2013) Pharmacologic properties, metabolism, and disposition of linaclotide, a novel therapeutic peptide approved for the treatment of irritable bowel syndrome with constipation and chronic idiopathic constipation. J Pharmacol Exp Ther 344:196–206
15. Sheppard DN, Welsh MJ (1999) Structure and function of the CFTR chloride channel. Physiol Rev 79:S23–S45
16. Strong TV, Boehm K, Collins FS (1994) Localization of cystic fibrosis transmembrane conductance regulator mRNA in the human gastrointestinal tract by in situ hybridization. J Clin Invest 93:347–354

17. de Lisle RC, Meldi L, Roach E, Flynn M et al (2009) Mast cells and gastrointestinal dysmotility in the cystic fibrosis mouse. PLoS ONE 4:e4283
18. Bodewes FA, Verkade HJ, Taminiau JA, Borowitz D et al (2015) Cystic fibrosis and the role of gastrointestinal outcome measures in the new era of therapeutic CFTR modulation. J Cyst Fibros 14:169–177
19. Lynch SV, Goldfarb KC, Wild YK, Kong W et al (2013) Cystic fibrosis transmembrane conductance regulator knockout mice exhibit aberrant gastrointestinal microbiota. Gut Microbes 4:41–47
20. Salaga M, Sobczak M, Fichna J (2013) Inhibition of proteases as a novel therapeutic strategy in the treatment of metabolic, inflammatory and functional diseases of the gastrointestinal tract. Drug Discov Today 18:708–715
21. Mosinska P, Zielinska M, Fichna J (2016) Expression and physiology of opioid receptors in the gastrointestinal tract. Curr Opin Endocrinol Diabetes Obes 23:3–10
22. Sobczak M, Salaga M, Storr MA, Fichna J. (2013) Physiology, signaling, and pharmacology of opioid receptors and their ligands in the gastrointestinal tract: current concepts and future perspectives. J Gastroenterol 49:24–45
23. Fichna J, Dicay M, Lewellyn K, Janecka A et al (2012) Salvinorin A has antiinflammatory and antinociceptive effects in experimental models of colitis in mice mediated by KOR and CB1 receptors. Inflamm Bowel Dis 18:1137–1145
24. Salaga M, Polepally PR, Zakrzewski PK, Cygankiewicz A et al (2014) Novel orally available salvinorin A analog PR-38 protects against experimental colitis and reduces abdominal pain in mice by interaction with opioid and cannabinoid receptors. Biochem Pharmacol 92:618–626
25. Salaga M, Polepally PR, Sobczak M, Grzywacz D et al (2014) Novel orally available salvinorin A analog PR-38 inhibits gastrointestinal motility and reduces abdominal pain in mouse models mimicking irritable bowel syndrome. J Pharmacol Exp Ther 350:69–78
26. Bokic T, Storr M, Schicho R (2015) Potential causes and present pharmacotherapy of irritable bowel syndrome: an overview. Pharmacology 96:76–85
27. Chang L, Ameen VZ, Dukes GE, McSorley DJ et al (2005) A dose-ranging, phase II study of the efficacy and safety of alosetron in men with diarrhea-predominant IBS. Am J Gastroenterol 100:115–123
28. Krause R, Ameen V, Gordon SH, West M et al (2007) A randomized, double-blind, placebo-controlled study to assess efficacy and safety of 0.5 mg and 1 mg alosetron in women with severe diarrhea-predominant IBS. Am J Gastroenterol 102:1709–1719
29. Tong K, Nicandro JP, Shringarpure R, Chuang E et al (2013) A 9-year evaluation of temporal trends in alosetron postmarketing safety under the risk management program. Therap Adv Gastroenterol 6:344–357
30. Chiba T, Yamamoto K, Sato S, Suzuki K (2013) Long-term efficacy and safety of ramosetron in the treatment of diarrhea-predominant irritable bowel syndrome. Clin Exp Gastroenterol 6:123–128
31. Matsueda K, Harasawa S, Hongo M, Hiwatashi N et al (2008) A phase II trial of the novel serotonin type 3 receptor antagonist ramosetron in Japanese male and female patients with diarrhea-predominant irritable bowel syndrome. Digestion 77:225–235
32. Matsueda K, Harasawa S, Hongo M, Hiwatashi N et al (2008) A randomized, double-blind, placebo-controlled clinical trial of the effectiveness of the novel serotonin type 3 receptor antagonist ramosetron in both male and female Japanese patients with diarrhea-predominant irritable bowel syndrome. Scand J Gastroenterol 43:1202–1211
33. Min YW, Rhee PL (2015) The clinical potential of ramosetron in the treatment of irritable bowel syndrome with diarrhea (IBS-D). Therap Adv Gastroenterol 8:136–142
34. Nee J, Zakari M, Lembo AJ (2015) Current and emerging drug options in the treatment of diarrhea predominant irritable bowel syndrome. Expert Opin Pharmacother 16:2781–2792
35. Lavo B, Stenstam M, Nielsen AL (1987) Loperamide in treatment of irritable bowel syndrome—A double-blind placebo controlled study. Scand J Gastroenterol Suppl 130:77–80

36. Karabulut GS, Beser OF, Erginoz E, Kutlu T et al (2013) The incidence of irritable bowel syndrome in children using the Rome III criteria and the effect of trimebutine treatment. J Neurogastroenterol Motil 19:90–93
37. Delvaux M, Wingate D (1997) Trimebutine: mechanism of action, effects on gastrointestinal function and clinical results. J Int Med Res 25:225–246
38. Fujita W, Gomes I, Dove LS, Prohaska D et al (2014) Molecular characterization of eluxadoline as a potential ligand targeting mu-delta opioid receptor heteromers. Biochem Pharmacol 92:448–456
39. Dove LS, Lembo A, Randall CW, Fogel R et al (2013) Eluxadoline benefits patients with irritable bowel syndrome with diarrhea in a phase 2 study. Gastroenterology 145:329–338
40. Lacy BE (2016) Emerging treatments in neurogastroenterology: eluxadoline—A new therapeutic option for diarrhea-predominant IBS. Neurogastroenterol Motil 28:26–35
41. Laterza L, Ianiro G, Scoleri I, Landi R et al (2015) Rifaximin for the treatment of diarrhoea-predominant irritable bowel syndrome. Expert Opin Pharmacother 16:607–615
42. Soldi S, Vasileiadis S, Uggeri F, Campanale M et al (2015) Modulation of the gut microbiota composition by rifaximin in non-constipated irritable bowel syndrome patients: a molecular approach. Clin Exp Gastroenterol 8:309–325
43. Kane JS, Ford AC (2016) Rifaximin for the treatment of diarrhea-predominant irritable bowel syndrome. Expert Rev Gastroenterol Hepatol 10:431–442
44. Pimentel M (2016) Review article: potential mechanisms of action of rifaximin in the management of irritable bowel syndrome with diarrhoea. Aliment Pharmacol Ther 43(Suppl 1):37–49
45. Yeo QM, Crutchley R, Cottreau J, Tucker A et al (2013) Crofelemer, a novel antisecretory agent approved for the treatment of HIV-associated diarrhea. Drugs Today (Barc) 49:239–252
46. Mangel AW, Chaturvedi P (2008) Evaluation of crofelemer in the treatment of diarrhea-predominant irritable bowel syndrome patients. Digestion 78:180–186
47. Chen CH, Lin CL, Kao CH (2016) Irritable bowel syndrome is associated with an increased risk of dementia: a nationwide population-based study. PLoS ONE 11:e0144589
48. Fichna J, Storr MA (2012) Brain-Gut Interactions in IBS. Front Pharmacol 3:127
49. Ford AC, Quigley EM, Lacy BE, Lembo AJ et al (2014) Effect of antidepressants and psychological therapies, including hypnotherapy, in irritable bowel syndrome: systematic review and meta-analysis. Am J Gastroenterol 109:1350–1365
50. Vahedi H, Merat S, Momtahen S, Kazzazi AS et al (2008) Clinical trial: the effect of amitriptyline in patients with diarrhoea-predominant irritable bowel syndrome. Aliment Pharmacol Ther 27:678–684
51. Awad RA, Camacho S (2010) A randomized, double-blind, placebo-controlled trial of polyethylene glycol effects on fasting and postprandial rectal sensitivity and symptoms in hypersensitive constipation-predominant irritable bowel syndrome. Colorectal Dis 12:1131–1138
52. Chapman RW, Stanghellini V, Geraint M, Halphen M (2013) Randomized clinical trial: macrogol/PEG 3350 plus electrolytes for treatment of patients with constipation associated with irritable bowel syndrome. Am J Gastroenterol 108:1508–1515
53. Tack J, Camilleri M, Chang L, Chey WD et al (2012) Systematic review: cardiovascular safety profile of 5-HT(4) agonists developed for gastrointestinal disorders. Aliment Pharmacol Ther 35:745–767
54. Brandt LJ (2008) The FDA's decision-making process: isn't it time to temper the principle of protective paternalism? Am J Gastroenterol 103:1226–1227
55. Chey WD, Pare P, Viegas A, Ligozio G et al (2008) Tegaserod for female patients suffering from IBS with mixed bowel habits or constipation: a randomized controlled trial. Am J Gastroenterol 103:1217–1225
56. Tougas G, Snape WJ Jr, Otten MH, Earnest DL et al (2002) Long-term safety of tegaserod in patients with constipation-predominant irritable bowel syndrome. Aliment Pharmacol Ther 16:1701–1708

57. Camilleri M, Kerstens R, Rykx A, Vandeplassche L (2008) A placebo-controlled trial of prucalopride for severe chronic constipation. N Engl J Med 358:2344–2354
58. Quigley EM, Vandeplassche L, Kerstens R, Ausma J (2009) Clinical trial: the efficacy, impact on quality of life, and safety and tolerability of prucalopride in severe chronic constipation—A 12-week, randomized, double-blind, placebo-controlled study. Aliment Pharmacol Ther 29:315–328
59. Tack J, van Beyens G, Kerstens R et al (2009) Prucalopride (Resolor) in the treatment of severe chronic constipation in patients dissatisfied with laxatives. Gut 58:357–365
60. Mugie SM, Korczowski B, Bodi P, Green A et al (2014) Prucalopride is no more effective than placebo for children with functional constipation. Gastroenterology 147:1285–1295
61. Johnston JM, Kurtz CB, Drossman DA, Lembo AJ et al (2009) Pilot study on the effect of linaclotide in patients with chronic constipation. Am J Gastroenterol 104:125–132
62. Rao S, Lembo AJ, Shiff SJ, Lavins BJ et al (2012) A 12-week, randomized, controlled trial with a 4-week randomized withdrawal period to evaluate the efficacy and safety of linaclotide in irritable bowel syndrome with constipation. Am J Gastroenterol 107:1714–1724
63. Chey WD, Lembo AJ, Lavins BJ, Shiff SJ et al (2012) Linaclotide for irritable bowel syndrome with constipation: a 26-week, randomized, double-blind, placebo-controlled trial to evaluate efficacy and safety. Am J Gastroenterol 107:1702–1712
64. Yu SW, Rao SS (2014) Advances in the management of constipation-predominant irritable bowel syndrome: the role of linaclotide. Therap Adv Gastroenterol 7:193–205
65. Wilson N, Schey R (2015) Lubiprostone in constipation: clinical evidence and place in therapy. Ther Adv Chronic Dis 6:40–50
66. Jakab RL, Collaco AM, Ameen NA (2012) Lubiprostone targets prostanoid signaling and promotes ion transporter trafficking, mucus exocytosis, and contractility. Dig Dis Sci 57:2826–2845
67. Johanson JF, Ueno R (2007) Lubiprostone, a locally acting chloride channel activator, in adult patients with chronic constipation: a double-blind, placebo-controlled, dose-ranging study to evaluate efficacy and safety. Aliment Pharmacol Ther 25:1351–1361
68. Johanson JF, Morton D, Geenen J, Ueno R (2008) Multicenter, 4-week, double-blind, randomized, placebo-controlled trial of lubiprostone, a locally-acting type-2 chloride channel activator, in patients with chronic constipation. Am J Gastroenterol 103:170–177
69. Johanson JF, Drossman DA, Panas R, Wahle A et al (2008) Clinical trial: phase 2 study of lubiprostone for irritable bowel syndrome with constipation. Aliment Pharmacol Ther 27:685–696
70. Trinkley KE, Nahata MC (2014) Medication management of irritable bowel syndrome. Digestion 89:253–267
71. Mosinska P, Salaga M, Fichna J (2016) Novel investigational drugs for constipation-predominant irritable bowel syndrome: a review. Expert Opin Investig Drugs 25:275–286
72. Deiana S, Gabbani T, Bagnoli S, Annese V (2015) Emerging drug for diarrhea predominant irritable bowel syndrome. Expert Opin Emerg Drugs 20:247–261
73. Alam MS, Roy PK, Miah AR, Mollick SH et al (2013) Efficacy of Peppermint oil in diarrhea predominant. Mymensingh Med J 22:27–30
74. Khanna R, Macdonald JK, Levesque BG (2014) Peppermint oil for the treatment of irritable bowel syndrome: a systematic review and meta-analysis. J Clin Gastroenterol 48:505–512
75. Ma XP, Hong J, An CP, Zhang D et al (2014) Acupuncture-moxibustion in treating irritable bowel syndrome: how does it work? World J Gastroenterol 20:6044–6054
76. Manheimer E, Wieland LS, Cheng K, Li SM et al (2012) Acupuncture for irritable bowel syndrome: systematic review and meta-analysis. Am J Gastroenterol 107:835–847
77. Stuardi T, MacPherson H (2012) Acupuncture for irritable bowel syndrome: diagnosis and treatment of patients in a pragmatic trial. J Altern Complement Med 18:1021–1027
78. Miller V, Carruthers HR, Morris J, Hasan SS et al (2015) Hypnotherapy for irritable bowel syndrome: an audit of one thousand adult patients. Aliment Pharmacol Ther 41:844–855
79. Palsson OS (2015) Hypnosis treatment of gastrointestinal disorders: A comprehensive review of the empirical evidence. Am J Clin Hypn 58:134–158

Chapter 5
IBS Patient's Guide

Marta Zielińska

Abstract In this chapter, we will discuss the role of intestinal microbiota, lifestyle and eating habits in IBS patients. Before pharmacological treatment patients should consider changing lifestyle and diet, as they may be potential triggers for IBS symptoms. We will indicate which daily products are proper and which should be avoided by IBS patients because of exacerbation of disease symptoms. We will shortly describe how to lead healthy life and what the impact of physical activity on IBS symptoms is. We will explain that control, but not treatment, is the key management concept in IBS patients. Finally, we will define psychological aspects of IBS development and how important is to maintain psychological homeostasis. The continuous control of the doctor and regular good contact between the doctor and the patient play an important role in the disease remission.

Keywords Diet · Fodmap · Lifestyle · Intestinal microbiota · Physical activity

Abbreviations

FODMAP	Fermentable oligosaccharides, disaccharides, monosaccharides, and polyol
GI	Gastrointestinal
IBS	Irritable bowel syndrome
IBS-C	Constipation-predominant-irritable bowel syndrome
IBS-D	Diarrhea predominant-irritable bowel syndrome
SIBO	Small intestinal bacterial overgrowth

M. Zielińska (✉)
Faculty of Medicine, Department of Biochemistry, Medical University of Lodz, Mazowiecka 6/8, 92-215 Lodz, Poland
e-mail: zielinska.martt@gmail.com

J. Fichna (ed.), *Introduction to gastrointestinal diseases Vol. 1*,
DOI 10.1007/978-3-319-49016-8_5

5.1 Intestinal Microflora

In the GI tract, we have more bacteria than cells in our entire body. Up to 1000 different microorganisms colonize the intestinal tract making about 2 kg of the body weight [1]. *Firmicutes* and *Bacteroidetes* are the major beneficial bacteria found in the GI tract of healthy people. Intestinal microbiome is specific for each individual and influenced by the genetic and environmental factors. The type and number of microbiota are also depended on—among others—age, gender and geographical origin.

Intestinal microbiota play many important roles in human health. They participate in food digestion, drug metabolism, detoxification and vitamin production. Intestinal microbiota regulate gut permeability and motility. Moreover, they have an impact on the integrity of the GI mucosa, immunomodulation (through prevention of pathogen colonization) and visceral sensitivity [2].

The changes in the intestinal microbiome are combined with the development of post-infectious IBS subtype. *Campylobacter*, *Escherichia coli*, *Salmonella*, *Shigella* and *Yersinia* are bacterial species involved in the development of infectious enterocolitis or gastroenteritis [3]. The prevalence of IBS after bacterial infection is 6–7 times higher than in patients without an infection episode [4]. The severity of infection increases the risk of disease development, for example diarrhea which lasts more than 7 or 21 day is associated with 2- or 3-fold higher risk of IBS development, respectively. Furthermore, the risk of IBS development is elevated for at least two years after infection.

Little is known about small intestine microbiome and its involvement in the course of IBS, mainly due to unavailability of small intestine tissues for basic research [5]. The small intestine contains a much lower density of bacteria than the colon in healthy subjects. Consequently, the small intestinal bacterial overgrowth (SIBO) may have an impact on IBS development and disease syndrome exacerbation. SIBO is a disorder of excessive growth of bacteria in the jejunum, which are typical in the colon. SIBO causes malabsorption and digestion problems. Patients with SIBO may have an increased amount of gases inside the intestine, what may result in abdominal pain, bloating and altered bowel function [6]. Chung et al. [7] found that IBS patients had higher abundance of *Veillonellaceae* and *Prevotellaceae*, but lower abundance of *Mycobacteriaceae* and *Neisseriaceae* in the small intestine. The ratio of *Firmicutes* to *Actinobacter* in IBS-A patients was approx. 9-fold (17.42/2.0 %) higher than in controls. Moreover, the ratio of *Firmicutes* to *Bacteroidetes* in the stool of IBS-D patients was 3-fold higher than healthy volunteers. In contrary, in a Swedish study no difference in small intestinal microbiota between IBS patients and healthy subjects was evidenced [8].

Further research is still needed to include or exclude the role of altered small intestinal microbiome in IBS. The differences in small intestinal microbiota found in many studies may result from geographical origin or the diagnostic tests used. Unfortunately, small intestinal microbiota can be measured only indirectly—by glucose hydrogen and lactulose hydrogen breath tests [9]. Currently available diagnostic tools should be improved and more standardized (for example: differences occur in the content of microbiota in the ileal effluent even in the same patient during the day) or novel assays should be engineered.

More is known about microbiota in the colon, where the studies are mainly based on the analysis of fecal samples. Luminal bacteria participate in digestion and modulation of the host immune system through their metabolites. *Bifidobacteria* and *Lactobacilli* are the major beneficial microbiota in the colon and in IBS patients a decreased number of these bacteria in comparison to healthy volunteers has been noted [10]. The decrease in *Bifidobacteria* and *Lactobacilli* impairs the GI homeostasis and may cause the mucosal inflammation in the GI tract [11]. There is also a difference in the abundance of *Enterobacteriaceae* species in fecal samples from IBS patients. For example, the populations of *Eubacterium—Clostridium coccoides* group, *Coprococcus, Collinsella, and Coprobacillus* species are altered in IBS patients [12]. The strains of *Veillonella* spp. are significantly increased in IBS-C patients [13].

There is a growing evidence that altered microbiome in the intestinal mucosa may also be implicated in IBS development. Mucosal bacterial make up differs from that of the fecal microbiome. The major role of the intestinal mucosa is to maintain a non-inflammatory state despite the presence of numerous microbiota in the intestines. In the colonic mucosa, the bacterial composition in IBS patients varies from healthy controls, for example an increased number of invasive bacteria, e.g. *Pseudomonas aeruginosa* and *Campylobacter jejuni* was noted.

Food intake also changes the composition of intestinal microbiota [14]. For example, the protein and animal fat consumption have been associated with the domination of *Bacteroides* enterotype, while a high carbohydrate intake was combined with the increase of *Prevotella* enterotype [15].

The other way to modulate gut microbiota is application of probiotics, prebiotics, synbiotics and antibiotics. All these are potential therapeutic options in IBS.

Probiotics are live microorganisms intended to provide health benefit for the host [16]. Probiotics include *Bifidobacterium* and *Lactobacillus*, lactic acid bacteria (*Lactococcus*, *Streptococcus*), organisms of the genera *Bacillus*, *Bacteroides* and *Enterococcus* [17]. The use of probiotics such as *Bifidobacterium* spp. and *Lactobacillus* spp. has been shown to have a positive effect on IBS symptoms [18]. Probiotics stimulate goblet cells to mucus production, what reduces visceral hypersensitivity, enhances the intestinal barrier function and normalizes bowel movements [19]. It was showed that probiotics exert beneficial effects to the host; however, they can act as a double-edged sword with both negative and positive effects. Therefore, precaution is necessary before the probiotic administration and during their long term usage the patient should be under control of the doctor [20].

Prebiotics are typically non-digestible carbohydrates, for example oligosaccharides: inulin, fructo-oligosaccharides, galacto-oligosaccharides and lactulose, which are fermented by bacteria with potential benefit to the host. Prebiotics affect mainly *Lactobacilli* and *Bifidobacteria*, because of selective stimulation of their growth [17]. Consumption of a certain prebiotic, trans-galactooligosaccharide for 4 weeks improved IBS symptoms and increased fecal *Bifidobacterium* spp. and

Eubacterium rectal/Clostridium coccoides ratio. Moreover, this prebiotic decreased the proportion of the *Clostridium perfringens–hystolyticum* and *Bacteroides/Prevotella* spp [21.] Prebiotics are present in leeks, asparagus, garlic, artichoke, onions, wheat, bananas, oats, and soy beans [22]. The consumption of prebiotics increases tolerance for high FODMAP food and adding regular exercise improves the beneficial effects of such diet [23].

Synbiotics are a mixture of selected probiotic strains and compatible prebiotics. For example, a synbiotic containing *Lactobacillus paracasei* and a prebiotic mixture improved the number of bowel movements, abdominal pain and IBS score in IBS-D patients. Well-being was also improved [24]. Another symbiotic mixture, containing lyophilised bacteria (*Lactobacilli*, *Bifidobacteria* and *Streptococcus thermophilus*) and inulin has shown a beneficial effect in alleviation of flatulence severity in IBS patients, but it failed to achieve an improvement in abdominal bloating [25].

Antibiotics are used in IBS therapy to treat imbalanced intestinal microbiota. Neomycin and rifaximin were broadly tested in clinical trials in IBS therapy. Neomycin was more effective than placebo in reducing IBS symptoms, but its action was combined with numerous side effects. There was a significant reduction in abdominal pain, dysfunctional defecation, bloating and abdominal discomfort in those IBS patients who received rifaximin compared to placebo. However, repeated administration of antibiotics in IBS therapy still remains controversial and should be under continuous control of the doctor.

Fecal microbial transplantation (FMT) is a novel approach to modulation of the gut microbiota, particularly in dysbiosis [26]. The role of FMT is reintroduction and re-establishment of a stable community of microbiota from a healthy donor to IBS patient. The scientific data about FMT are confusing, therefore it is not a standard anti-IBS therapy and it still needs to be improved.

5.2 Diet and Lifestyle

5.2.1 Diet

Diet is important in our daily life, we should choose appropriate food products consciously, because of their continuous impact on our health. Starting the day with big healthy breakfast, which should the most important meal gives a lot of energy for the whole day. We should also take care about lunch—forget about processed food and snacks. Dinner should be eaten not just before going to bed, but reasonably early (at least 3 h earlier). Finally, we should use to drink two liters of water per day. We know it all, but reality is different. Remember, not only food consumption, but also irregular and improper eating habits represent an important issue in our diet. Eating is not only about satisfying hunger—meals should not be eaten in a hurry—but constitute a part of the day (consumption with friends or family, not alone or in front of TV).

Diet has still not been proven as a cause of IBS or implicated in disease development. However, there are many clinical studies indicating that adequate diet may have an impact on attenuation of IBS symptoms—mainly abdominal pain, disturbances in GI motility and flatulence, which are significantly involved in decreased quality of life in IBS patients [27]. IBS patients had significantly more irregular meal habits and skipped meals (which caused a loss in gastro-colonic reflex and restrains defecation) than healthy individuals [28, 29].

It was found that IBS patients complained after certain food products; GI disturbances were reported within 15 min in 28 % and within 3 h in 93 % of patients [30]. Moreover, it was also revealed that 60 % of IBS patients exclude some food from diet, because of more severe disease symptoms [31]. Nanda et al. found that onions, garlic, paprika, beans, peas and chocolate are the most common food products incriminated by IBS patients [32, 33]. Also, an acute chili ingestion aggravated abdominal pain and burning symptoms of IBS [34]. Furthermore, rice and wheat have been combined to bloating and diarrhea [30]. In contrary, higher consumption of canned food, processed meat, confectionary, chocolate and herbal tea was noted in IBS patients. It was also revealed that mean intake of protein and salt was higher in IBS patients than the recommendations [35]. Interestingly, women reported more intolerable food items than men [36]. Women with IBS ate less fish, fruit, milk, and green-yellow vegetables than men with IBS and healthy individuals [29]. All these data strongly support the hypothesis that better understanding of food intake and dietary management may constitute a tool for controlling IBS course [37].

A first step to improve IBS symptoms using non-pharmacological tools is the avoidance of fat and highly processed food. Fast food, potato chips, popcorn and fried foods may interfere with the intestinal movements and may result in symptoms such as constipation and diarrhea. After high fat meal IBS patients more frequently complain about fullness, bloating and nausea than healthy people.

Next step is the reduction of alcohol and caffeine consumption, because they were found to have an impact on abdominal pain or discomfort, bloating or change in bowel habit for at least 6 months [38].

The traditional American and European diet is not rich in fiber, therefore many physicians advise fiber supplements for abdominal pain reduction and altered GI motility in IBS patients. Fiber is any food which is not absorbed and broken down through the GI tract. There are two types of fiber: soluble (present in whole grains, wheat born) and insoluble depending on their interaction with water, and further classified into highly, intermediate, minimally or non-fermentable fiber (present in dried beans, peas).

Fiber starts to be digested in the large intestine to short-chain fatty acids (SCFA) and gases. Probably through SCFA production fiber increases the luminal osmotic load, attracting water, and has an impact on the microbiome resulting in an increased biomass [39]. Consumption of food containing fiber causes changes in colonic pH, an increase in stool bulk, acceleration of the whole GI transit and decrease of intracolonic pressure [40]. Reduced amount of pressure that bowel uses to move intestinal content may also cause alleviation of abdominal pain. Fiber consumption is recommended mainly in IBS-C patients, but it can be also helpful in IBS-D patients to firm up loose stools.

Eating of 20–30 g of fiber per day is sufficient and is defined as high fiber diet. To increase the content of fiber in diet, a lot of fruit (banana, blueberry, figs, kiwi, mango, orange, cherry) should be eaten. Another suggestion is to add dried peas, beans, whole grains to starters, soups and main dishes. High fiber should be associated with increased drinking of water or other healthy drinks (smoothie, fresh juices).

For IBS patients, it is recommended to try diet rich in fiber food, but carefully. If too much and too often fiber food would be added to their diet, it can escalate symptoms of the disease. Flatulence, bloating and abdominal pain are the most frequent IBS symptoms, which can be affected by high fiber diet [40, 41]. Flatulence is an individual feature and it should be carefully observed by the patient if gas production increases after certain food. There are fruits and vegetables rich in fiber, which increase production of gases, therefore should be avoided by IBS patients (most common listed in Table 5.1). In contrary, meat, fowl, fish, rice are the products with high content of fiber, but which do not cause excess production of gas.

A new trend in diet in IBS patients is reduced consumption of fermentable oligosaccharides, disaccharides, monosaccharides, and polyol (FODMAP) [42]. FODMAP food include products with high amount of fructose (pears, apples), oligosaccharides including fructans (wheat and onion), galacto-oligosaccharides (legumes: kidney beans and chickpeas) and sugar polyols such as sorbitol, xylitol or mannitol (artificial sweeteners) [43, 44]. Almost all of the highly processed food (main dishes, fast food and sauces) contain FODMAP.

FODMAPs are present in grains, some dairy products - milk, sour cream (with lower content of fat), kefir, yogurt, butter, some cheeses. Onions, garlic, asparagus, beets, leeks, broccoli, cauliflower, Brussels sprouts, chicory, fennel are rich in FODMAPs. Peaches, avocados, nectarines, plums, cherries, watermelon, melon, blackberries, lychee, mango, guava, papaya, avocado contain FODMAPs in high concentrations. Moreover, honey and liqueur wines also include FODMAPs.

Lethargy, increased GI symptoms (bloating, abdominal pain, passage of wind and dissatisfaction with stool consistency) and higher levels of breath hydrogen are produced on high FODMAP diet [45].

Table 5.1 Fruits and vegetables with high fiber content

Gas-producing food with high fiber content		Less gas-producing food with high fiber content	
Vegetable	Fruit	Vegetable	Fruit
Broccoli	Apple	Apricot	Carrots
Brussels sprout	Grape	Pineapple	Corn
Cauliflower	Banana	Berries	Green
Cabbage	Raisin	Orange	Tomato
Cucumber	Prunes	Peach	Spinach

A low FODMAP diet could help decrease the distention caused by both the osmotic effect of FODMAPs and gas production resulting from its fermentation in the colon [46]. Moreover, lowering FODMAP consumption clearly reduced the relative abundance of all intestinal bacteria [27]. Of course, it is impossible to rule out FODMAPs from diet, but all these benefits indicate that it should be under consideration of IBS patients to minimize the FODMAPs consumption. Citrus fruit (oranges, lemons, limes, grapes) and forest fruit (cranberries, blueberries, raspberries and strawberries) are suitable for IBS patients on low FODMAPs diet. Vegetables which can be consumed are potatoes, peppers, carrots, cucumbers, zucchini, tomatoes, radishes, sweet potatoes, bamboo sprouts, olives, Chinese cabbage, and lettuce. Thyme, rosemary, basil, ginger, mint and oregano are herbs and spices which should enrich main dishes. Fruit and vegetables with high and low FODMAP content are listed in Table 5.2.

Gluten-free food should be introduced to the diet instead of wheat products. Wheat could be successfully replaced by spelt, which is known to contain fewer galactans and fructans than wheat and therefore not to produce frequent IBS symptoms [47]. It was evidenced that gluten-free diet improved IBS symptoms [48]. Patients with IBS-D, who received gluten-free diet (bread and muffin without gluten) reported a significant improvement in the following symptoms: pain, bloating, stool consistency, and tiredness as compared to IBS-D patients who ingested gluten (bread and muffin, 16 g gluten per day) [49].

IBS patients often complain due to lactase deficiency. Lactase is an enzyme involved in digestion of lactose—sugar in milk. The most common symptoms of this ailment include cramping abdominal pain, bloating, flatulence, diarrhea and nausea. IBS patients with lactase intolerance should avoid high-lactose food: dairy products: milk, sour cream, cheese (also cottage cheese, ricotta, spread cheese) and ice cream. However, they should remember that dairy products are a big source of calcium, potassium, magnesium, vitamin A, vitamin B2, vitamin B12 and other microelements and therefore they risk the development of these nutrient deficiencies. Vitamins and microelements should be replaced in other food products or supplemented [50].

Fructose malabsorption should be considered in the handling of patients with IBS complaints. Fructose reduced diet should result in lower fructose intake (less than 2 g per meal) and allow IBS symptoms improvement [51].

Table 5.2 Fruit and vegetables with high and low FODMAP content

High FODMAP content		Low FODMAP content	
Vegetable	Fruit	Vegetable	Fruit
Asparagus	Apple	Carrot	Banana
Garlic	Pear	Celery	Raspberry
Cabbage	Mango	Lettuce	Strawberry
Onion	Watermelon	Corn	Orange
Pea	Nashi pear	Tomato	Grape

Box 5.1 Diet recommendations
What to avoid:

- overeating
- high-fiber food
- high-FODMAP food
- dairy products
- artificial sweeteners
- fried and processed food
- chocolate
- carbonated beverages
- alcohol
- caffeine
- white bread
- red meat
- spicy food
- onion

Recommendations:

- increase fluid intake
- drink warm or hot drinks
- drink small amounts of alcohol (only during dinner)
- drink a lot of herbal tea
- eat slowly and regularly
- eat low-FODMAP food
- eat low- to medium-fiber food
- forget about processed food
- eat dinner 3 h before sleep

5.2.2 *Obesity*

The correlation between IBS development and obesity is not clear and not confirmed in big scale clinical trials. Data are conflicting—in one study an association between low body mass index and IBS has been found [52], while in another study it was evidenced that most IBS patients are normal-weight or overweight [53]. High-fat diet has been shown to have an impact on the intestinal microbiota and thus may contribute to more severe IBS symptoms in obese patients.

5.2.3 *Alcohol Consumption*

The consumption of alcohol, mainly wine and beer, was also described as a factor involved in the exacerbation of disease symptoms, and therefore it should be avoided [54]. Finally, alcohol drinks with carbonated beverages (sweetened with mannitol or sorbitol) should also be excluded from diet, because they facilitate gas production [33].

5.2.4 *Social Life*

IBS is a chronic and relapsing disorder and its symptoms decrease patients' quality of life. Disease symptoms often complicate outgoing lifestyle of patients; patients often avoid friend appointments—especially in the restaurants. This uncertainty of when and where disease symptoms may occur can cause fear when patient is away from home. Moreover, they feel psychical discomfort because of lack of easy access to toilet during friend meetings [55].

5.2.5 *Physical Activity*

Active lifestyle and physical activity should be pivotal from early years to adult. People who practice sports are more conscientious as compared to inactive ones. Moreover, systemic trainings can help to maintain regular life style. There is an increased risk of IBS development in physically inactive people. It was noted that physically active IBS patients reported not so severe disease symptoms as compared with physically inactive patients [56]. For example, active women were less likely to report a feeling of incomplete evacuation than inactive ones. Moreover, daily exercise can help to maintain good intestinal function, prevent bloating and are effective in relieving constipation [57]. Finally, daily exercise can improve mood and symptoms of fatigue, which are also more frequently noted in IBS patients [58].

Yoga is recommended in IBS patients, because it combines physical postures, breathing exercises and meditation or relaxation. Yoga can have beneficial effects on the emotional and the physical symptoms of IBS, thus can help to cope with stress. However, yoga is safe only when practiced appropriately.

Not all the patients are satisfied after training—in some IBS patients strenuous exercise may act on the intestines as a stressor. Therefore it should be taken into consideration whether increased physical activity will help to alleviate or exacerbate IBS symptoms [59].

5.2.6 Sleep Disturbances

Sleep is a time needed for regeneration after day full of work; sleep and dream disturbances influence IBS symptoms [60]. For example, the history of being psychologically abused and less than 6 h of sleep are combined with more severe disease course and fear of symptoms exacerbation. The time of sleep for IBS patients should be longer than 6 h. Patients should also take care about quality of their dream, for example last caffeine beverage should be drunk 4 h before bed time. Moreover, an important issue is the maintenance of bed time frame (both in the evening and in the morning). Bed should be used only for sleeping or sexual activity, not for eating, watching TV or book reading. One of the possible solutions to improve quality of sleep is relaxation exercise or yoga. In IBS patients regular napping periods during the afternoon should be avoided since lethargy further aggravates IBS symptoms.

In conclusion, regular exercise, smoking cessation, abstinence from alcohol, and maintenance of regular eating habits can be easily achieved by IBS patients in daily life without their doctor's assistance and should be the first approach in IBS management.

5.3 Psychological Aspects and Treatment

The term "brain-gut axis" refers to the bi-directional communication between the gut (enteric nervous system) and the central nervous system. Brain-gut axis plays a crucial role in gut function modulation in health and disease (Fig. 5.1). The human *psyche* is affected by many factors, including personality features, altered health beliefs, coping skills and psychological factors. They all have impact on the motor, sensory, secretory and immune functions of the GI tract through the brain-gut axis [61].

Anger proneness and expression style may be associated with pro-inflammatory processes and visceral hypersensitivity that contribute to IBS signs and symptoms [62]. Patients with IBS had significantly higher levels of trait anger than healthy subjects [63]. The trait anger represents a stable dispositional feature and includes a general predisposition to become angry.

IBS symptoms have impact on daily function, thoughts, feelings and behaviors because of the impression that disease symptoms can be aggravated anytime. Moreover, IBS patients indicate that they lost sense of freedom, social contacts, but gained feelings of fearfulness and embarrassment due to frequent visits in the toilet [62].

Patients with IBS are more likely to be psychiatrically ill (panic, anxiety, mood disorders, depression and post-traumatic stress disorder) than the general population [35]. On the contrary, people who are more prone to fear, anxiety and affective disorders more frequently suffer from IBS symptoms [64]. Depression constitutes

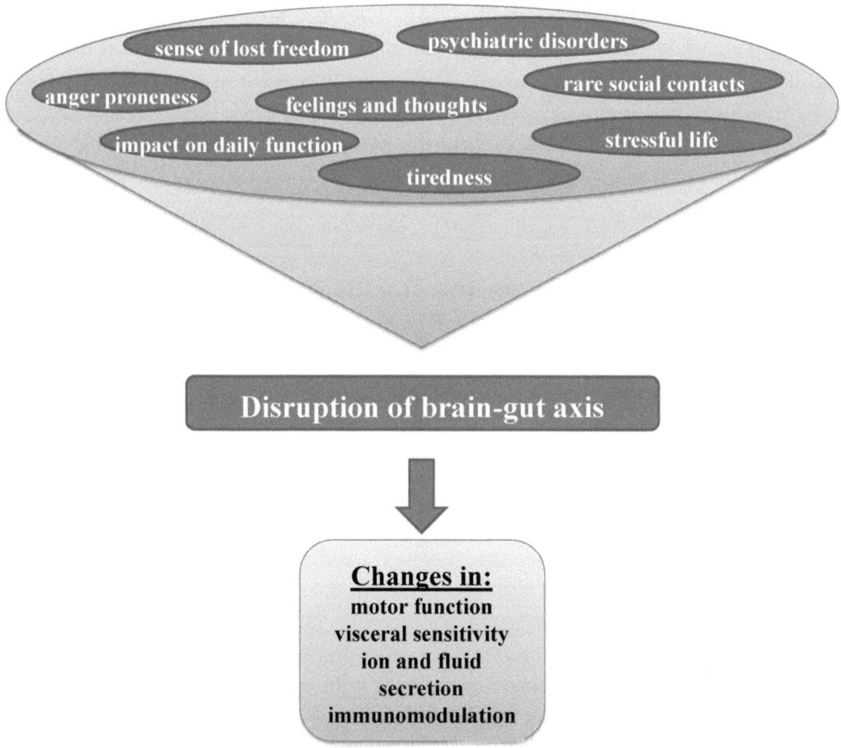

Fig.5.1 Brain-gut axis

risk factor for the development of IBS and is the most common psychiatric disorder diagnosed in IBS patients [65].

Social problems, tiredness, dizziness, excitedness, and excessive use of health care services (including alternative medicine) occur more frequently in IBS patients as compared to healthy individuals [66]. Moreover, worrisome and stressful life events have been reported to be associated with more severe IBS symptoms [67]. The major life events (divorce, unemployment, death of a relative) or social events (social changes, revolution) influence IBS [61]. There is apparent correlation between stress loading and exacerbation of GI symptoms in IBS patients [68]: when psychosocial stress was loaded on IBS patients in an examination room, GI transit was accelerated, as determined by measurement of colonic manometry [69].

Currently available therapeutics used in IBS therapy that target psychological disturbances include anxiolytic agents and antidepressants [70]. The mechanism of antidepressants action involves their participation in pain modulation (peripheral analgesic effect), improved quality of sleep, and regulation of GI motility [71].

Non-pharmacological forms of psychological treatments used in IBS therapy include psychotherapy (cognitive-behavioral therapy), relaxation therapy and hypnotherapy [72]. Gut-directed hypnotherapy improves IBS symptoms, mainly abdominal pain, and quality of life [73]. The mechanism through which hypnotherapy alleviates IBS symptoms is still unclear, but it was postulated that rectal sensitivity to distension is decreased. The major limitation of hypnotherapy is low number of qualified therapists and high costs of visits [74].

The choice of treatment depends on the patient requirements, available resources, and the experience of the doctor.

5.4 Co-operation Between the Doctor and the Patient

Only one third of IBS patients search for advice from a family physician or an internist. Most of IBS patients do not consider their symptoms serious enough to consult the doctor and try to lead own control of the disease and therapy. They often receive medical information from the Internet, brochures and books and from a nurse. Only when IBS symptoms are exacerbated, they look for help from the gastroenterologist.

Patients with IBS often think that they are insufficiently informed in relation to risk of serious GI diseases and the role of diet in the course of IBS [75]. They have a feeling that doctors do not listen to them or do not understand their illness experience. Moreover, IBS patients feel only partially satisfied with their information about disease as compared to patients with diabetes mellitus, hypertension or heart disease [76]. Consequently, a detailed and comprehensive explanation of the disease should be the first step in communication with IBS patient. Education is a very important part of IBS treatment—for example, it was evidenced that IBS patients who participated in psychoeducational program reported improvement in symptoms severity and quality of life [77].

After diagnosis, IBS patients should realize that IBS is a chronic incurable disorder and being under continuous control of the doctor is extremely important (even in relapsing periods of IBS), not only when they have symptoms exacerbations and need a quick help.

Doctor consultation should be the first line of choice in the management of IBS, mainly because of health professional's knowledge, experience and ability to notice other characteristics that IBS patient exhibit, e.g. anxiety or depression. Regular appointments with a doctor is a key to effective therapy in IBS. The information obtained during examination is on both sides—patient's and doctor's [78]. The patient should ask the doctor about all deliberations according to proper lifestyle without any embarrassment. The doctor should ask about disease symptoms and their severity, including frequency of defecations, relief after defecation, abnormal stool, blood in stool and presence of nausea or flatulence (Table 5.3).

Doctor can ask about a brief dietary history, any associated factors (like daily obligations, stressors, sleep disturbances, used drugs). It can help to identify dietary and/or other factors that may have an impact on disease course.

Doctor should be focused on patient concerns and expectations of therapy. Doctor should observe or ask about warning symptoms, such as unexplained weight

Table 5.3 Questions which need to be answered during the visit

Issues addressed during appointment	
Doctor's side	Patient's side
Do you feel satisfied with current drugs and treatment or you want to discuss it?	Could I have other GI disorders?
Were there any stressful situations since your last visit?	Do I lead a proper life style? Do you recommend any changes?
What about severity of symptoms? Improvement of exacerbation?	Can physical activity help in symptoms improvement?
Brief dietary history	Are there any other possible therapies?
Do you have problems with sleep?	What about traditional medicine?
Psychological aspects of IBS	Do I need psychiatric consultation?
Do you use regularly any other drugs?	Which food should I avoid?

loss, progressive or unrelenting pain, GI bleeding, longitudinal diarrhea during consultation. In particular, any patient older than 50 years of age should undergo a detailed examination to confirm the absence of a colon cancer.

The preservation of warmth and empathy between doctors and their patients will make an important contribution to improved quality of life of patients instead of a brief doctor visit (only for prescription). It has been also demonstrated that patients who see the same doctor during consecutive consultations are less anxious and simultaneously more satisfied with their treatment process.

Patient knows everything about his/her body, therefore can determine if the current therapy brings satisfying benefits. If they do not feel any improvement in health, possibilities of alternative treatments should be broadly discussed—including changing lifestyle, conventional treatments (e.g. suppositories, creams, heat pad) and alternative modalities (e.g. hypnotherapy, acupuncture, homeopathy).

Doctors should remind their patients that they should not forget to lead a normal social life and try not to think negatively about the disease. The doctor should ask about daily life—sport activities, sleep quality, stressful events. Moreover, the doctor should ensure the patient that social contacts are pivotal—patients should benefit from being with family and friends—not staying at home. Finally, IBS patients should take short holidays few times a year. Being outgoing and active seems to bring a lot of benefits for them.

As mentioned above, IBS is combined with brain-gut disturbances, and psychiatric diseases are more frequently noted in IBS patients. The doctor should observe IBS patient and react when any additional help from a psychiatrist is needed. Many doctors refer IBS patients to psychological and psychiatric clinics, but they sometimes do not realize that it may paradoxically further escalate patient's confusion and frustration. Sometimes it is better to just listen to the patient's needs.

In conclusion, most IBS patients benefit from a therapeutic relationship with the doctors. An experimentally applied supportive patient-doctor relationship significantly improved symptoms and quality of life in IBS [79]. The establishment of a positive patient-doctor relationship reduces the number of appointments (which should stay regular) and improves long-term therapy, although it has not been confirmed in any clinical trials, mainly because of the nature of the intervention [80, 81].

Acknowledgments Supported by grant from the Medical University of Lodz (#and 502-03/1-156-04/502-14-297) and grants from National Science Centre (#UMO-2013/11/N/NZ7/00724 and UMO-2014/12/T/NZ7/00252). MZ is the recipient of the Polish L'Oréal UNESCO Awards for Women in Science and Polpharma Foundation Scholarship.

References

1. Biedermann L, Rogler G (2015) The intestinal microbiota: its role in health and disease. Eur J Pediatr 174(2):151–167
2. Ohman L, Simren M (2013) Intestinal microbiota and its role in irritable bowel syndrome (IBS). Curr Gastroenterol Rep 15(5):323
3. Porter CK, Choi D, Cash B, Pimentel M, Murray J, May L et al (2013) Pathogen-specific risk of chronic gastrointestinal disorders following bacterial causes of foodborne illness. BMC Gastroenterol 13:46
4. Halvorson HA, Schlett CD, Riddle MS (2006) Postinfectious irritable bowel syndrome–a meta-analysis. Am J Gastroenterol 101(8):1894–1899
5. Lee BJ, Bak YT (2011) Irritable bowel syndrome, gut microbiota and probiotics. J Neurogastroenterol Motil 17(3):252–266
6. Ghoshal UC, Kumar S, Mehrotra M, Lakshmi C, Misra A (2010) Frequency of small intestinal bacterial overgrowth in patients with irritable bowel syndrome and chronic non-specific diarrhea. J Neurogastroenterol Motil 16(1):40–46
7. Chung CS, Chang PF, Liao CH, Lee TH, Chen Y, Lee YC et al (2016) Differences of microbiota in small bowel and faeces between irritable bowel syndrome patients and healthy subjects. Scand J Gastroenterol 51(4):410–419
8. Dlugosz A, Winckler B, Lundin E, Zakikhany K, Sandstrom G, Ye W et al (2015) No difference in small bowel microbiota between patients with irritable bowel syndrome and healthy controls. Sci Rep 5:8508
9. Posserud I, Stotzer PO, Bjornsson ES, Abrahamsson H, Simren M (2007) Small intestinal bacterial overgrowth in patients with irritable bowel syndrome. Gut 56(6):802–808
10. Patcharatrakul T, Gonlachanvit S (2016) Chili peppers, curcumins, and prebiotics in gastrointestinal health and disease. Curr Gastroenterol Rep 18(4):19
11. Kerckhoffs AP, Ben-Amor K, Samsom M, van der Rest ME, de VJ, Knol J et al. (2011) Molecular analysis of faecal and duodenal samples reveals significantly higher prevalence and numbers of Pseudomonas aeruginosa in irritable bowel syndrome. J Med Microbiol 60(Pt 2):236–245
12. Si JM, Yu YC, Fan YJ, Chen SJ (2004) Intestinal microecology and quality of life in irritable bowel syndrome patients. World J Gastroenterol 10(12):1802–1805
13. Salonen A, de Vos WM, Palva A (2010) Gastrointestinal microbiota in irritable bowel syndrome: present state and perspectives. Microbiology 156(Pt 11):3205–3215
14. Cuomo R, Andreozzi P, Zito FP, Passananti V, De CG, Sarnelli G (2014) Irritable bowel syndrome and food interaction. World J Gastroenterol 20(27):8837–8845
15. Wu GD, Chen J, Hoffmann C, Bittinger K, Chen YY, Keilbaugh SA et al (2011) Linking long-term dietary patterns with gut microbial enterotypes. Science 334(6052):105–108
16. Simren M, Barbara G, Flint HJ, Spiegel BM, Spiller RC, Vanner S et al (2013) Intestinal microbiota in functional bowel disorders: a Rome foundation report. Gut 62(1):159–176

17. Bonfrate L, Tack J, Grattagliano I, Cuomo R, Portincasa P (2013) Microbiota in health and irritable bowel syndrome: current knowledge, perspectives and therapeutic options. Scand J Gastroenterol 48(9):995–1009
18. Sisson G, Ayis S, Sherwood RA, Bjarnason I (2014) Randomised clinical trial: A liquid multi-strain probiotic versus placebo in the irritable bowel syndrome–a 12 week double-blind study. Aliment Pharmacol Ther 40(1):51–62
19. Gareau MG, Sherman PM, Walker WA (2010) Probiotics and the gut microbiota in intestinal health and disease. Nat Rev Gastroenterol Hepatol 7(9):503–514
20. Didari T, Mozaffari S, Nikfar S, Abdollahi M (2015) Effectiveness of probiotics in irritable bowel syndrome: updated systematic review with meta-analysis. World J Gastroenterol 21 (10):3072–3084
21. Silk DB, Davis A, Vulevic J, Tzortzis G, Gibson GR (2009) Clinical trial: the effects of a trans-galactooligosaccharide prebiotic on faecal microbiota and symptoms in irritable bowel syndrome. Aliment Pharmacol Ther 29(5):508–518
22. Roberfroid M, Gibson GR, Hoyles L, McCartney AL, Rastall R, Rowland I et al (2010) Prebiotic effects: metabolic and health benefits. Br J Nutr 104(Suppl 2):S1–S63
23. Mazzawi T, Hausken T, Gundersen D, El-Salhy M (2013) Effects of dietary guidance on the symptoms, quality of life and habitual dietary intake of patients with irritable bowel syndrome. Mol Med Rep 8(3):845–852
24. Ford AC, Quigley EM, Lacy BE, Lembo AJ, Saito YA, Schiller LR et al (2014) Efficacy of prebiotics, probiotics, and synbiotics in irritable bowel syndrome and chronic idiopathic constipation: systematic review and meta-analysis. Am J Gastroenterol 109(10):1547–1561
25. Bogovic MB, Obermajer T, Lipoglavsek L, Sernel T, Locatelli I, Kos M et al (2016) Effects of synbiotic fermented milk containing Lactobacillus acidophilus La-5 and Bifidobacterium animalis ssp. lactis BB-12 on the fecal microbiota of adults with irritable bowel syndrome: A randomized double-blind, placebo-controlled trial. J Dairy Sci 99(7):5008–5021
26. Borody TJ, Khoruts A (2012) Fecal microbiota transplantation and emerging applications. Nat Rev Gastroenterol Hepatol 9(2):88–96
27. Iacovou M, Tan V, Muir JG, Gibson PR (2015) The low FODMAP diet and its application in East and Southeast Asia. J Neurogastroenterol Motil 21(4):459–470
28. Miwa H (2012) Life style in persons with functional gastrointestinal disorders–large-scale internet survey of lifestyle in Japan. Neurogastroenterol Motil 24(5):464–471 (e217)
29. Okami Y, Kato T, Nin G, Harada K, Aoi W, Wada S et al (2011) Lifestyle and psychological factors related to irritable bowel syndrome in nursing and medical school students. J Gastroenterol 46(12):1403–1410
30. Simren M, Mansson A, Langkilde AM, Svedlund J, Abrahamsson H, Bengtsson U et al (2001) Food-related gastrointestinal symptoms in the irritable bowel syndrome. Digestion 63 (2):108–115
31. Bohn L, Storsrud S, Tornblom H, Bengtsson U, Simren M (2013) Self-reported food-related gastrointestinal symptoms in IBS are common and associated with more severe symptoms and reduced quality of life. Am J Gastroenterol 108(5):634–641
32. Nanda R, James R, Smith H, Dudley CR, Jewell DP (1989) Food intolerance and the irritable bowel syndrome. Gut 30(8):1099–1104
33. Mazzawi T, Hausken T, Gundersen D, El-Salhy M (2013) Effects of dietary guidance on the symptoms, quality of life and habitual dietary intake of patients with irritable bowel syndrome. Mol Med Rep 8(3):845–852
34. Gonlachanvit S (2010) Are rice and spicy diet good for functional gastrointestinal disorders? J Neurogastroenterol Motil 16(2):131–138
35. Omagari K, Murayama T, Tanaka Y, Yoshikawa C, Inoue S, Ichimura M et al (2013) Mental, physical, dietary, and nutritional effects on irritable bowel syndrome in young Japanese women. Intern Med 52(12):1295–1301
36. Bohn L, Storsrud S, Tornblom H, Bengtsson U, Simren M (2013) Self-reported food-related gastrointestinal symptoms in IBS are common and associated with more severe symptoms and reduced quality of life. Am J Gastroenterol 108(5):634–641

37. El-Salhy M, Gundersen D (2015) Diet in irritable bowel syndrome. Nutr J 14:36
38. Staudacher HM, Whelan K, Irving PM, Lomer MC (2011) Comparison of symptom response following advice for a diet low in fermentable carbohydrates (FODMAPs) versus standard dietary advice in patients with irritable bowel syndrome. J Hum Nutr Diet 24(5):487–495
39. Gonlachanvit S, Coleski R, Owyang C, Hasler W (2004) Inhibitory actions of a high fibre diet on intestinal gas transit in healthy volunteers. Gut 53(11):1577–1582
40. Eswaran S, Muir J, Chey WD (2013) Fiber and functional gastrointestinal disorders. Am J Gastroenterol 108(5):718–727
41. Shepherd SJ, Parker FC, Muir JG, Gibson PR (2008) Dietary triggers of abdominal symptoms in patients with irritable bowel syndrome: randomized placebo-controlled evidence. Clin Gastroenterol Hepatol 6(7):765–771
42. Halmos EP, Power VA, Shepherd SJ, Gibson PR, Muir JG (2014) A diet low in FODMAPs reduces symptoms of irritable bowel syndrome. Gastroenterology 146(1):67–75
43. Gibson PR, Shepherd SJ (2005) Personal view: food for thought–western lifestyle and susceptibility to Crohn's disease. The FODMAP hypothesis. Aliment Pharmacol Ther 21 (12):1399–1409
44. Marsh A, Eslick EM, Eslick GD (2016) Does a diet low in FODMAPs reduce symptoms associated with functional gastrointestinal disorders? A comprehensive systematic review and meta-analysis. Eur J Nutr 55(3):897–906
45. Staudacher HM, Lomer MC, Anderson JL, Barrett JS, Muir JG, Irving PM et al (2012) Fermentable carbohydrate restriction reduces luminal bifidobacteria and gastrointestinal symptoms in patients with irritable bowel syndrome. J Nutr 142(8):1510–1518
46. Ong DK, Mitchell SB, Barrett JS, Shepherd SJ, Irving PM, Biesiekierski JR et al (2010) Manipulation of dietary short chain carbohydrates alters the pattern of gas production and genesis of symptoms in irritable bowel syndrome. J Gastroenterol Hepatol 25(8):1366–1373
47. Williams EA, Nai X, Corfe BM (2011) Dietary intakes in people with irritable bowel syndrome. BMC Gastroenterol 11:9
48. Biesiekierski JR, Muir JG, Gibson PR (2013) Is gluten a cause of gastrointestinal symptoms in people without celiac disease? Curr Allergy Asthma Rep 13(6):631–638
49. Biesiekierski JR, Newnham ED, Irving PM, Barrett JS, Haines M, Doecke JD et al (2011) Gluten causes gastrointestinal symptoms in subjects without celiac disease: a double-blind randomized placebo-controlled trial. Am J Gastroenterol 106(3):508–514
50. Wald A, Rakel D (2008) Behavioral and complementary approaches for the treatment of irritable bowel syndrome. Nutr Clin Pract 23(3):284–292
51. Berg LK, Fagerli E, Martinussen M, Myhre AO, Florholmen J, Goll R (2013) Effect of fructose-reduced diet in patients with irritable bowel syndrome, and its correlation to a standard fructose breath test. Scand J Gastroenterol 48(8):936–943
52. Kubo M, Fujiwara Y, Shiba M, Kohata Y, Yamagami H, Tanigawa T et al (2011) Differences between risk factors among irritable bowel syndrome subtypes in Japanese adults. Neurogastroenterol Motil 23(3):249–254
53. Simren M, Mansson A, Langkilde AM, Svedlund J, Abrahamsson H, Bengtsson U et al (2001) Food-related gastrointestinal symptoms in the irritable bowel syndrome. Digestion 63 (2):108–115
54. Ostgaard H, Hausken T, Gundersen D, El-Salhy M (2012) Diet and effects of diet management on quality of life and symptoms in patients with irritable bowel syndrome. Mol Med Rep 5(6):1382–1390
55. Bertram S, Kurland M, Lydick E, Locke GR III, Yawn BP (2001) The patient's perspective of irritable bowel syndrome. J Fam Pract 50(6):521–525
56. Guo YB, Zhuang KM, Kuang L, Zhan Q, Wang XF, Liu SD (2015) Association between diet and lifestyle habits and irritable bowel syndrome: a case-control study. Gut Liver 9(5):649–656
57. Lustyk MK, Jarrett ME, Bennett JC, Heitkemper MM (2001) Does a physically active lifestyle improve symptoms in women with irritable bowel syndrome? Gastroenterol Nurs 24 (3):129–137

58. Sullivan SN, Wong C, Heidenheim P (1994) Does running cause gastrointestinal symptoms? A survey of 93 randomly selected runners compared with controls. N Z Med J 107(984):328–331
59. Levy RL, Cain KC, Jarrett M, Heitkemper MM (1997) The relationship between daily life stress and gastrointestinal symptoms in women with irritable bowel syndrome. J Behav Med 20(2):177–193
60. Fujii Y, Nomura S (2008) A prospective study of the psychobehavioral factors responsible for a change from non-patient irritable bowel syndrome to IBS patient status. Biopsychosoc Med 2:16
61. Padhy SK, Sahoo S, Mahajan S, Sinha SK (2015) Irritable bowel syndrome: Is it "irritable brain" or "irritable bowel"? J Neurosci Rural Pract 6(4):568–577
62. Surdea-Blaga T, Baban A, Dumitrascu DL (2012) Psychosocial determinants of irritable bowel syndrome. World J Gastroenterol 18(7):616–626
63. Stanculete MF, Pojoga C, Dumitrascu DL (2014) Experience of anger in patients with irritable bowel syndrome in Romania. Clujul Med 87(2):98–101
64. Tosic-Golubovic S, Miljkovic S, Nagorni A, Lazarevic D, Nikolic G (2010) Irritable bowel syndrome, anxiety, depression and personality characteristics. Psychiatr Danub 22(3):418–424
65. Koloski NA, Jones M, Kalantar J, Weltman M, Zaguirre J, Talley NJ (2012) The brain–gut pathway in functional gastrointestinal disorders is bidirectional: a 12-year prospective population-based study. Gut 61(9):1284–1290
66. Donker GA, Foets M, Spreeuwenberg P (1999) Patients with irritable bowel syndrome: health status and use of health care services. Br J Gen Pract 49(447):787–792
67. Song SW, Park SJ, Kim SH, Kang SG (2012) Relationship between irritable bowel syndrome, worry and stress in adolescent girls. J Korean Med Sci 27(11):1398–1404
68. Whitehead WE, Crowell MD, Robinson JC, Heller BR, Schuster MM (1992) Effects of stressful life events on bowel symptoms: subjects with irritable bowel syndrome compared with subjects without bowel dysfunction. Gut 33(6):825–830
69. Fukudo S, Suzuki J (1987) Colonic motility, autonomic function, and gastrointestinal hormones under psychological stress on irritable bowel syndrome. Tohoku J Exp Med 151 (4):373–385
70. Chang FY (2014) Irritable bowel syndrome: the evolution of multi-dimensional looking and multidisciplinary treatments. World J Gastroenterol 20(10):2499–2514
71. Clouse RE, Lustman PJ (2005) Use of psychopharmacological agents for functional gastrointestinal disorders. Gut 54(9):1332–1341
72. Spiller R, Aziz Q, Creed F, Emmanuel A, Houghton L, Hungin P et al (2007) Guidelines on the irritable bowel syndrome: mechanisms and practical management. Gut 56(12):1770–1798
73. Ford AC, Talley NJ, Schoenfeld PS, Quigley EM, Moayyedi P (2009) Efficacy of antidepressants and psychological therapies in irritable bowel syndrome: systematic review and meta-analysis. Gut 58(3):367–378
74. Flik CE, van Rood YR, Laan W, Smout AJ, Weusten BL, Whorwell PJ et al (2011) A randomised controlled trial on hypnotherapy for irritable bowel syndrome: design and methodological challenges (the IMAGINE study). BMC Gastroenterol 11:137
75. O'Sullivan MA, Mahmud N, Kelleher DP, Lovett E, O'Morain CA (2000) Patient knowledge and educational needs in irritable bowel syndrome. Eur J Gastroenterol Hepatol 12(1):39–43
76. Halpert A, Dalton CB, Palsson O, Morris C, Hu Y, Bangdiwala S et al (2008) Patient educational media preferences for information about irritable bowel syndrome (IBS). Dig Dis Sci 53(12):3184–3190
77. Ringstrom G, Storsrud S, Posserud I, Lundqvist S, Westman B, Simren M (2010) Structured patient education is superior to written information in the management of patients with irritable bowel syndrome: a randomized controlled study. Eur J Gastroenterol Hepatol 22(4):420–428
78. Di Palma JA, Herrera JL (2012) The role of effective clinician-patient communication in the management of irritable bowel syndrome and chronic constipation. J Clin Gastroenterol 46 (9):748–751
79. Halpert A (2011) Irritable bowel syndrome: what do patients really want? Curr Gastroenterol Rep 13(4):331–335

80. Kelley JM, Lembo AJ, Ablon JS, Villanueva JJ, Conboy LA, Levy R et al (2009) Patient and practitioner influences on the placebo effect in irritable bowel syndrome. Psychosom Med 71 (7):789–797
81. Vanuytsel T, Tack JF, Boeckxstaens GE (2014) Treatment of abdominal pain in irritable bowel syndrome. J Gastroenterol 49(8):1193–1205

Part II
Inflammatory Bowel Disease

Chapter 6
Pathogenesis of IBD

Aleksandra Sobolewska-Włodarczyk and Marcin Włodarczyk

Abstract Crohn's disease (CD) and ulcerative colitis (UC) are the main representatives of inflammatory bowel diseases (IBD). IBD are defined as a group of chronic, immune system-mediated inflammatory diseases of the gastrointestinal (GI) tract (Xavier and Podolsky in Nature 448:427–434, 2007 [1]). The pathogenesis of IBD is not fully understood; however, a similar cytokine activation profile is observed in psoriasis, rheumatoid arthritis and systemic lupus erythematosus, which are all associated with generalized immune imbalance (Mikhailov and Furner in World J Gastroenterol 15(3):270–279, 2009 [2]; Baumgart and Carding in Lancet 369 (9573):1627–1640, 2007 [3]; Kaser et al. in Annu Rev Immunol 28:573–621, 2010 [4]). On the other hand, clinical symptoms differ among these diseases and may involve various organs. Importantly, environmental and infectious factors, together with genetic predisposition lead to elevated levels of pro-inflammatory cytokines and specific (abnormal) tissue responses during the course of IBD (Podolsky in N Engl J Med 347(6):417–429, 2002 [5]; Molodecky et al. in Gastroenterology 142 (1):46–54, 2012 [6]). Recent studies suggest that the etiology of IBD involves environmental and genetic factors that cause dysfunction of the epithelial barrier with consequent deregulation of the mucosal immune system and responses to gut microbiota. In this chapter, an overview to IBD pathogenesis will be presented.

A. Sobolewska-Włodarczyk (✉) · M. Włodarczyk (✉)
Faculty of Medicine, Department of Biochemistry, Medical University of Lodz, Mazowiecka 6/8, 92-215 Lodz, Poland
e-mail: olasobolewska1@poczta.onet.pl

M. Włodarczyk
e-mail: dr.mwlodarczyk@gmail.com

A. Sobolewska-Włodarczyk
Faculty of Military Medicine, Department of Gastroenterology, Medical University of Lodz, Lodz, Poland

M. Włodarczyk
Faculty of Military Medicine, Department of General and Colorectal Surgery, Medical University of Lodz, Hallera Sq 1, Lodz, Poland

J. Fichna (ed.), *Introduction to gastrointestinal diseases Vol. 1*,
DOI 10.1007/978-3-319-49016-8_6

Keywords Inflammatory bowel disease · Pathogenesis · Crohn's disease · Ulcerative colitis · Risk factors

The chart below shows the current view of the development of IBD.

6.1 Dysfunction of the Intestinal Barrier

Histologically, the intestinal wall has four layers:

(a) mucosa,
(b) submucosa,
(c) muscular layer,
(d) serosa.

Mucosa with epithelial cells constitute a specific form of a "link with the outside world" and their role is to form a physical, chemical, and immune barrier. The epithelium, the most exposed part of the mucosa, is a glandular barrier with goblet cells that forms the luminal surface. Goblet cells secrete mucus, which lubricates the passage of food along and protects epithelium from digestive enzymes. In healthy individuals, the mucus layer protects the epithelium and the layers below from luminal bacteria and allows interactions mainly through the Peyer's patches. Any diminished mucosal protection may lead to an increased bacterial adhesion and invasion, with a final inflammatory process. Disturbance in the mucosal barrier seems to be the key element in the onset of IBD and, subsequently, in the frequent relapses [7]. Alterations in the small vasculature of the mucosal layer, followed by the appearance of aphthous ulcers is the earliest pathological and endoscopic step in the course of IBD [8].

6.2 Immunological Reaction

The immune system plays a key role in the development of IBD. The implicated cells include intestinal epithelia, innate lymphoid cells, macrophages, dendritic cells, B cells, and T cells. The interaction of the antigen-presenting cells (APCs) with bacterial antigens leads to differentiation of naïve T-cells into effector T-helper cells, which occurs mainly in Peyer's patches and lymphoid tissue. These reactions lead to immunological imbalance and overproduction of proinflammatory cytokines, especially interleukins and tumor necrosis factor-alpha.

Cytokines are a broad and loose category of small proteins (~5–20 kDa) that are important in cell signaling. They are released by cells and affect the behavior of other cells. Cytokines can also be involved in autocrine signaling. The group of cytokines includes chemokines, interferons, interleukins, lymphokines, and tumor necrosis factors, but generally no hormones or growth factors (despite some overlap in the terminology).

Interleukins are a group of cytokines that were first seen to be expressed by white blood cells (leukocytes) [9]. The function of the immune system depends in large part on interleukins, and any irregularities in their turnover usually results in autoimmune diseases or immune deficiency.

Tumor necrosis factor alpha (TNF-α) is the best-known member of the cytokine subdivision called apoptosis cytokines. TNF-α is a monocyte-derived cytokine that has been implicated in tumor regression, septic shock, and cachexia [10, 11]. The protein is synthesized as a prohormone with an unusually long and atypical signal sequence, which is absent from the mature secreted cytokine [12].

In IBD, cytokine imbalance leads to a chronic intestinal inflammation. Clinical manifestations and inflammatory lesions in the intestinal wall are induced by elevated levels of several cytokines, mainly TNF-α. Recent studies point at other than TNF-α related pathways in the pathogenesis of IBD, in particular in CD, and suggest a strong link between IL-23, IL-17A, TNF-α and interferon γ (IFN-γ) [13, 14]. In a study by Hovhannisyan et al. [15], elevated levels of IL-17A were observed in the mucosa and serum of CD patients. This may suggest that in some cases IBD development may be due to an excessive activation of TNF-α pathway and simultaneous Th17 lymphocyte activation dependent on IL-23. Another study demonstrated that $CD4^+$ Th17 lymphocytes are responsible for skin lesions and gut inflammation in IBD [16].

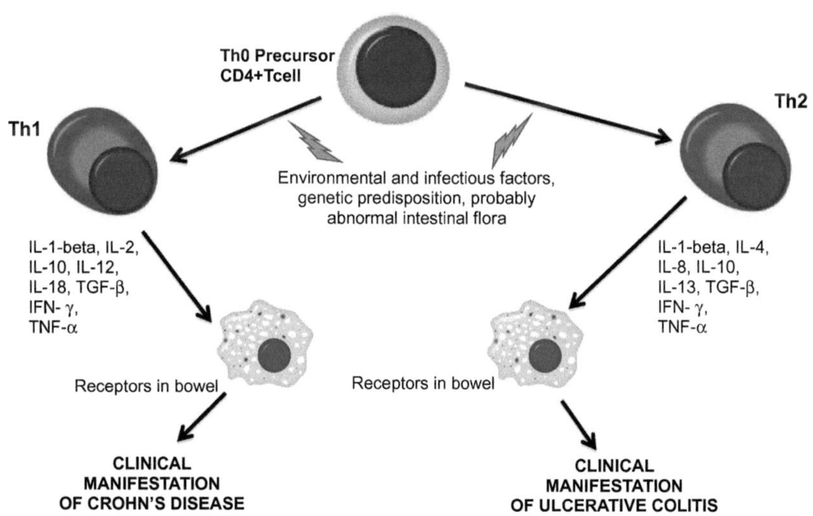

Fig. 6.1 The abnormal activation of T helper (Th)1 and Th2 cells in the development of IBD

In case of CD the imbalance between proinflammatory and anti-inflammatory cytokines leads to a disproportionate activation of T helper (Th)1 cells and overproduction of interleukin (IL)-1β, IL-2, IL-10, IL-12, IL-18, transforming growth factor beta (TGF-β) and tumor necrosis factor-α (TNF-α) [13, 14]. The UC development seems to be Th2-dependent and the immune-mediated process leads to overproduction IL-4, IL-5, IL-6, IL-10, and IL-13 (Fig. 6.1).

6.3 Risk Factors

6.3.1 Genetic Factors

The first proof for possible genetic basis in CD was provided by the studies on monozygotic twins and other familial clusters of IBD [17]. Lately, genome-wide association studies (GWAS) identified more than 150 genetic risk loci for IBD, 70 of which may be associated with CD. Recent, a great variation between European, American, and Asian populations was shown, with different gene mutations that can predispose to IBD, especially to CD [18].

The first CD GWAS was conducted in Japanese, in 2005, and identified tumor necrosis factor superfamily member 15 (TNFSF15) as a susceptibility locus (genetic variant, which increases the probability of contracting the disease but is not 'necessary' or 'sufficient' for disease expression) [17]. This was followed by a rush of studies from 2006 to 2008 [18–27], each including approximately 500–2000 CD cases and a similar number of controls genotyped at 100,000–600,000 single nucleotide polymorphisms (SNPs).

One of the most important associations in IBD pathogenesis is the polymorphism in the nucleotide-binding oligomerization domain-containing protein 2 (NOD2)/ caspase recruitment domain-containing protein 15 (CARD15) gene. NOD2 is a protein in the NF-kB pathway and in humans it is encoded by the NOD2 gene located on chromosome 16 [28, 29]. NOD2 plays an important role in the immune system: it recognizes bacterial molecules (peptidoglycans) and stimulates an immune reaction, and acts as an intracellular sensor for bacterial wall components, especially muramyl dipeptide. Clinically, variants of NOD2 are associated with ileal involvement, a stenosing or fistulizing pattern of disease and a higher risk of surgery [30].

Other genetic variants that may lead to an increased risk of IBD are: toll-like receptor 4 (TLR-4), caspase recruitment domain-containing protein 9 (CARD9), interleukin 23 receptor (IL-23R), signal transducer and activator of transcription 3 (STAT3) for innate immunity, human leukocyte antigen, interferon regulatory factor 5 (IRF-5), protein tyrosine phosphatase non-receptor type 22 (PTPN-22) for adaptive immune system, etc. [18].

Recently, genetic variations in the autophagic pathway were shown to be the risk factors in IBD pathogenesis. Autophagy is a lysosomal recycling mechanism of the cytoplasm that plays an important role in the innate immune response toward intracellular bacteria. Yet another pathway with close interaction to the autophagic pathway is the unfolded protein response induced by endoplasmic reticulum stress. Autophagy-related 16-like 1 gene (ATG16L1) and immunity-related guanosine triphosphatase gene (IRGM) have been linked to a higher susceptibility of CD [31].

Micro-RNAs (miRNAs) usually contain about 22 nucleotides. These are small non-coding RNA molecules found in plants, animals and some viruses, that function in RNA silencing and post-transcriptional regulation of gene expression [32, 33]. Over 5400 miRNAs have been identified so far, each carrying possible implications in autoimmune-mediated diseases [18]. In the context of IBD, correlations to the NOD-like receptors, TLRs and T-helper cells, especially Th17, were mentioned [34]. Because these are only observational studies, their therapeutic application is not entirely clear. The limitations of these molecules are represented by difficulty to target a specific organ and the associated systemic adverse reactions. However, miRNAs depending on the type of IBD could become a biomarker of disease in the near future [35].

6.3.2 *Environmental Factors*

Long-term observations evidenced different frequency of IBD depending on the geographical region. Most importantly, higher occurrence of IBD was observed in highly developed countries, what suggests—among others—that low exposure to pathogenic infections could disturb the mucosal immune balance, thus increasing the risk of IBD [36]. Previous or present smoking has been typically associated with a higher risk of CD [37]. Nicotine is related to an increased epithelial cell apoptosis, a higher intestinal permeability, and also to changes in the mucosal immune response without a clearly proven correlation. Tobacco smoke constituents could also influence the intestinal immune balance by lowering T-cell proliferation and altering macrophagic response [38].

In the past smoking was noticed as protective factor of UC. In 2016, To et al. [39] conducted a meta-analysis about the effect of tobacco smoking on the natural history of UC. The study showed that smoking does not improve the natural history of UC. Given the health benefits of smoking cessation and the lack of clear benefit in UC, smoking cessation advice should thus be incorporated into guidance on the management of the disease.

Diet is considered a pathological trigger in some cases, as feeding habits can affect intestinal permeability and efficient clearance of bacterial antigens, consequently influencing the immune system [40]. In a recent study, Kawaguchi et al. [41] showed that food antigens can trigger $CD4^+$ T cell activation in the mouse model of CD and are associated with high IgG plasma levels. With the help of GWAS correlated with nutrigenetic and nutrigenomic research, a new, more personalized approach to the patient with IBD is expected [41, 42]. There are clear evidences that nutritional therapy is highly successful in the treatment of CD. Exclusive enteral nutrition is well established as remission induction therapy. New diets, such as a CD exclusion diet or defined diets (specific carbohydrate diets, FODMAP diet, Paleolithic diet) are currently being discussed as treatment options for IBD patients [43].

6.3.3 Microbial Factors

6.3.3.1 Adherent-Invasive *E. Coli*

Interest in *Escherichia coli* as a pathogen in IBD began when it was shown that microorganisms isolated from patients with CD had greater adherent properties to human cells than those from controls, and that previously unrecognized invasive *E. coli* were present in Crohn's ileal tissue [44–46].

Darfeuille-Michaud et al. [45] reported that *E. coli* was recovered from 65 % of chronic lesions in resected ileum and 100 % of biopsies of early lesions in post-operative endoscopic recurrence. Recent studies showed that *E. coli* strains are able to adhere to various human cells or cell lines. Wine et al. showed that 53–62 % of *E. coli* strains isolated from feces of CD were able to adhere to buccal cells, compared to only 5–6 % of those isolated from control subjects. In the same study, the correlation between bacterial adhesion to intestinal cells and intestinal colonization has been observed. The presence of high levels of bacteria creates a biofilm on the surface of the gut mucosa in patients with CD and UC [47, 48].

Glasser et al. [49] demonstrated that adherent-invasive *Escherichia coli* (AIEC) was able to survive and replicate in macrophages, without inducing host cells and stimulating the infected cells to release high levels of TNF-α.

6.3.3.2 Mycobacterium Avium Paratuberculosis

Mycobacterium avium subspecies paratuberculosis (MAP) is a pathogenic microorganism that causes Johne's disease in ruminants and other animals such as primates and rabbits [50]. Because of clinical similarities between Johne's disease in ruminants and IBD in humans, some researchers point at MAP as the cause of CD [44]. In line, similarly to CD MAP infection causes segmental and fibrosing stenosis, as well as epithelial granulomata [51]. Of note, in 1913 Dalziel et al. showed a correlation between CD and MAP [52]. Consequently, Naser et al. [53] cultured MAP from blood in up to 50 % of CD patients and 22 % of UC patients, but no control subjects. Curiously, antibiotic treatment against MAP does not cure IBD.

6.3.3.3 Helicobacter Pylori

Helicobacter pylori (HP) infection may be a protective factor against chronic inflammatory diseases, like IBD. A large study conducted by Väre et al. [54]

confirmed the low prevalence of HP infection, especially in CD patients and showed that the age of onset of IBD was higher in seropositive than in seronegative patients.

In 2010, a meta-analysis that evaluated the possible relationship between IBD and HP infection was published. Luther et al. [55] showed that 27 % of the IBD patients had HP infection, in comparison to 40 % of the control group, with an estimated relative risk of infection of 0.64 %. The authors suggested a protective role of HP infection in IBD pathogenesis, but it was also noted that several heterogeneous factors could influence the study results [55].

In 2011, the same research group published a paper that tried to clarify the mechanism responsible for the inverse association of HP and IBD. The authors postulated that HP DNA in distal intestine could influence mucosal immunity. They also showed that it is capable of inhibiting the production of proinflammatory cytokines of the murine or human cells in vitro [56]. However, other studies have exposed that this lower prevalence of HP in IBD may be secondary to HP "spontaneous eradication" with 5-ASA or antibiotic treatment [57].

6.3.3.4 Clostridium Difficile

Clostridium difficile (C. difficile) is a gram positive bacillus and may become established in the human colon. C. difficile is present in 2–5 % of the adult population [58]. From time to time antibiotic therapy, especially clindamycin has the adverse effect of disrupting the normal balance of the gut flora, in which case C. difficile may opportunistically dominate, causing C. difficile colitis with watery diarrhea. Recent studies showed that in 40 % of the cases in IBD patients C. difficile colitis can appear without previous use of antibiotic [59, 60].

In 2013, Nitzan et al. [61] published an extensive review regarding the role of C. difficile in the pathogenesis of IBD, as well as its implications with respect to diagnosis and treatment. The review embraces different risk factors, clinical characteristics of the infection in IBD, special aspects of its presentation, diagnosis and treatment in IBD. It was noted that C. difficile infection in IBD patients most likely plays a role in the pathogenesis of exacerbations, although probably not in the development of IBD itself [61].

6.3.3.5 Viruses

Based on epidemiology studies, two theories about the relationship between viral infections and the development of IBD have been proposed. The first theory suggests that certain infections that occur during infancy may predispose to the appearance of IBD. The second theory, Hygiene Theory, points to the absence of infections in infancy and the lack of contact with certain antigens as the cause of subsequent intestinal inflammation. Several epidemiological studies about coincidence of viral infection and IBD development were conducted. Consequently, measles, mumps, cytomegalovirus, virus de Epstein-Barr were connected to IBD [62–66]. Nowadays rather implication in exacerbation, earlier age of IBD onset and viruses infections neither in IBD development is postulated.

Acknowledgments Supported by the National Science Center (2015/17/N/NZ5/00677 to ASW).

References

1. Xavier RJ, Podolsky DK (2007) Unravelling the pathogenesis of inflammatory bowel disease. Nature 448(7152):427–434
2. Mikhailov TA, Furner SE (2009) Breastfeeding and genetic factors in the etiology of inflammatory bowel disease in children. World J Gastroenterol 15(3):270–279
3. Baumgart DC, Carding SR (2007) Inflammatory bowel disease: cause and immunobiology. Lancet 369(9573):1627–1640
4. Kaser A, Zeissig S, Blumberg RS (2010) Inflammatory bowel disease. Annu Rev Immunol 28:573–621
5. Podolsky DK (2002) Inflammatory bowel disease. N Engl J Med 347(6):417–429
6. Molodecky NA, Soon IS, Rabi DM et al (2012) Increasing incidence and prevalence of the inflammatory bowel diseases with time, based on systematic review. Gastroenterology 142 (1):46–54
7. Antoni L, Nuding S, Wehkamp J, Stange EF (2014) Intestinal barrier in inflammatory bowel disease. World J Gastroenterol 20(5):1165–1179
8. Sankey EA, Dhillon AP, Anthony A et al (1993) Early mucosal changes in Crohn's disease. Gut 34(3):375–381
9. "Cytokine" in John Lackie (2010) A dictionary of biomedicine. Oxford University Press. ISBN 9780199549351
10. Fransen L, Müller R, Marmenout A, Tavernier J, Van der Heyden J, Kawashima E, Chollet A, Tizard R, Van Heuverswyn H, Van Vliet A (1985) Molecular cloning of mouse tumour necrosis factor cDNA and its eukaryotic expression. Nucleic Acids Res 13(12):4417–4429
11. Kriegler M, Perez C, DeFay K, Albert I, Lu SD (1988) A novel form of TNF/cachectin is a cell surface cytotoxic transmembrane protein: ramifications for the complex physiology of TNF. Cell 53(1):45–53
12. Sherry B, Jue DM, Zentella A, Cerami A (1990) Characterization of high molecular weight glycosylated forms of murine tumor necrosis factor. Biochem Biophys Res Commun 173 (3):1072–1078
13. Cobrin GM, Abreu MT (2005) Defects in mucosal immunity leading to Crohn's disease. Immunol Rev 206:277–295
14. Targan SR, Karp LC (2005) Defects in mucosal immunity leading to ulcerative colitis. Immunol Rev 206:296–305
15. Hovhannisyan Z, Treatman J, Littman DR, Mayer L (2011) Characterization of interleukin-17-producing regulatory T cells in inflamed intestinal mucosa from patients with inflammatory bowel diseases. Gastroenterology 140(3):957–965. doi:10.1053/j.gastro.2010.12.002 Epub 2010 Dec 11
16. Włodarczyk M, Sobolewska A, Wójcik B, Loga K, Fichna J, Wiśniewska-Jarosińska M (2014) Correlations between skin lesions induced by anti-tumor necrosis factor-α and selected cytokines in Crohn's disease patients. World J Gastroenterol 20(22):7019–7026
17. Van Limbergen J, Wilson DC, Satsangi J (2009) The genetics of Crohn's disease. Annu Rev Genomics Hum Genet 10:89–116
18. Liu JZ, Anderson CA (2014) Genetic studies of Crohn's disease: past, present and future. Best Pract Res Clin Gastroenterol 28(3):373–386

19. Yamazaki K, McGovern D, Ragoussis J, Paolucci M, Butler H, Jewell D (2005) Single nucleotide polymorphisms in TNFSF15 confer susceptibility to Crohn's disease. Hum Mol Genet 14:3499–3506
20. Hampe J, Franke A, Rosenstiel P, Till A, Teuber M, Huse K (2007) A genome-wide association scan of nonsynonymous SNPs identifies a susceptibility variant for Crohn disease in ATG16L1. Nat Genet 39:207–211
21. Duerr RH, Taylor KD, Brant SR, Rioux JD, Silverberg MS, Daly MJ (2006) A genome-wide association study identifies IL23R as an inflammatory bowel disease gene. Science 314: 1461–1463
22. Rioux JD, Xavier RJ, Taylor KD, Silverberg MS, Goyette P, Huett A (2007) Genome-wide association study identifies new susceptibility loci for Crohn disease and implicates autophagy in disease pathogenesis. Nat Genet 39:596–604
23. Libioulle C, Louis E, Hansoul S, Sandor C, Farnir F, Franchimont D (2007) Novel Crohn disease locus identified by genome-wide association maps to a gene desert on 5p13. 1 and modulates expression of PTGER4. PLoS Genet 3:e58
24. Parkes M, Barrett JC, Prescott NJ, Tremelling M, Anderson CA, Fisher SA (2007) Sequence variants in the autophagy gene IRGM and multiple other replicating loci contribute to Crohn's disease susceptibility. Nat Genet 39:830–832
25. Franke A, Hampe J, Rosenstiel P, Becker C, Wagner F, Häsler R (2007) Systematic association mapping identifies NELL1 as a Novel IBD Disease Gene. PLoS ONE 2:e691
26. Raelson JV, Little RD, Ruether A, Fournier H, Paquin B, Van Eerdewegh P (2007) Genome-wide association study for Crohn's disease in the Quebec founder population identifies multiple validated disease loci. Proc Natl Acad Sci USA 104:14747–14752
27. Control Wellcome Trust Case (2007) C Genome-wide association study of 14,000 cases of seven common diseases and 3000 shared controls. Nature 447:661–678
28. Gilberts EC, Greenstein AJ, Katsel P, Harpaz N, Greenstein RJ (1994) Molecular evidence for two forms of Crohn disease. Proc Natl Acad Sci USA 91(26):12721–12724
29. Hugot JP, Laurent-Puig P, Gower-Rousseau C, Olson JM, Lee JC, Beaugerie L, Naom I, Dupas JL, Van Gossum A, Orholm M, Bonaiti-Pellie C, Weissenbach J, Mathew CG, Lennard-Jones JE, Cortot A, Colombel JF, Thomas G (1996) Mapping of a susceptibility locus for Crohn's disease on chromosome 16. Nature 379(6568):821–823
30. Adler J, Rangwalla SC, Dwamena BA, Higgins PD (2011) The prognostic power of the NOD2 genotype for complicated Crohn's disease: a meta-analysis. Am J Gastroenterol 106(4):699–712
31. Fritz T, Niederreiter L, Adolph T, Blumberg RS, Kaser A (2011) Crohn's disease: NOD2, autophagy and ER stress converge. Gut 60(11):1580–1588
32. Ambros V (2004) The functions of animal microRNAs. Nature 431(7006):350–355
33. Bartel DP (2004) MicroRNAs: genomics, biogenesis, mechanism, and function. Cell 116(2):281–297
34. Archanioti P, Gazouli M, Theodoropoulos G, Vaiopoulou A, Nikiteas N (2011) Micro-RNAs as regulators and possible diagnostic bio-markers in inflammatory bowel disease. J Crohns Colitis 5(6):520–524
35. Chapman CG, Pekow J (2015) The emerging role of miRNAs in inflammatory bowel disease: a review. Therap Adv Gastroenterol 8(1):4–22
36. Sartor RB (2006) Mechanisms of disease: pathogenesis of Crohn's disease and ulcerative colitis. Nat Clin Pract Gastroenterol Hepatol 3(7):390–407
37. Slattery E, Mitchell P, Mulcahy HE (2011) Cigarette smoking in Crohns disease: can we do more? J Crohns Colitis 5(5):505
38. Graham C (2013) Understanding smoking and nicotine effects on the gastrointestinal tract. Gastrointest Nurs 11(1):44
39. To N, Ford AC, Gracie DJ (2016) Systematic review with meta-analysis: the effect of tobacco smoking on the natural history of ulcerative colitis. Aliment Pharmacol Ther 44(2):117–126
40. Pfeffer-Gik T, Levine A (2014) Dietary clues to the pathogenesis of Crohn's disease. Dig Dis 32(4):389–394

41. Kawaguchi T, Mori M, Saito K et al (2014) Food antigen-induced immune responses in Crohn's disease patients and experimental colitis mice. J Gastroenterol 50(4):394–405
42. Ferguson LR (2012) Potential value of nutrigenomics in Crohn's disease. Nat Rev Gastroenterol Hepatol 9(5):260–270
43. Ruemmele FM (2016) Role of diet in inflammatory bowel disease. Ann Nutr Metab 68(Suppl 1):33–41
44. Hansen R, Thomson JM, El-Omar EM, Hold GL (2010) The role of infection in the aetiology of inflammatory bowel disease. J Gastroenterol 45:266–276
45. Darfeuille-Michaud A, Neut C, Barnich N, Lederman E, Di Martino P, Desreumaux P, Gambiez L, Joly B, Cortot A, Colombel JF (1998) Presence of adherent *Escherichia coli* strains in ileal mucosa of patients with Crohn's disease. Gastroenterology 115:1405–1413
46. O'Brien CL, Pavli P, Gordon DM, Allison GE (2014) Detection of bacterial DNA in lymph nodes of Crohn's disease patients using high throughput sequencing. Gut 63:1596–1606
47. Barnich N, Darfeuille-Michaud A (2007) Role of bacteria in the etiopathogenesis of inflammatory bowel disease. World J Gastroenterol 13:5571–5576
48. Wine E, Ossa JC, Gray-Owen SD, Sherman PM (2010) Adherent-invasive *Escherichia coli* target the epithelial barrier. Gut Microbes 1:80–84
49. Glasser AL, Boudeau J, Barnich N, Perruchot MH, Colombel JF, Darfeuille-Michaud A (2001) Adherent invasive *Escherichia coli* strains from patients with Crohn's disease survive and replicate within macrophages without inducing host cell death. Infect Immun 69:5529–5537
50. Friswell M, Campbell B, Rhodes J (2010) The role of bacteria in the pathogenesis of inflammatory bowel disease. Gut Liver 4:295–306
51. Lidar M, Langevitz P, Shoenfeld Y (2009) The role of infection in inflammatory bowel disease: initiation, exacerbation and protection. Isr Med Assoc J 11:558–563
52. Hermon-Taylor J, Moss M, Tizard M, Malik Z, Sanderson J (1990) Molecular biology of Crohn's disease mycobacteria. Baillieres Clin Gastroenterol 4(1):23–42
53. Naser SA, Ghobrial G, Romero C, Valentine JF (2004) Culture of Mycobacterium avium subspecies paratuberculosis from the blood of patients with Crohn's disease. Lancet 364:1039–1044
54. Väre PO, Heikius B, Silvennoinen JA, Karttunen R, Niemelä SE, Lehtola JK, Karttunen TJ (2001) Seroprevalence of Helicobacter pylori infection in inflammatory bowel disease: is Helicobacter pylori infection a protective factor? Scand J Gastroenterol 36:1295–1300
55. Luther J, Dave M, Higgins PD, Kao JY (2010) Association between Helicobacter pylori infection and inflammatory bowel disease: a meta-analysis and systematic review of the literature. Inflamm Bowel Dis 16:1077–1084
56. Luther J, Owyang SY, Takeuchi T, Cole TS, Zhang M, Liu M, Erb-Downward J, Rubenstein JH, Chen CC, Pierzchala AV et al (2011) Helicobacter pylori DNA decreases pro-inflammatory cytokine production by dendritic cells and attenuates dextran sodium sulphate-induced colitis. Gut 60:1479–1486
57. Papamichael K, Konstantopoulos P, Mantzaris GJ (2014) Helicobacter pylori infection and inflammatory bowel disease: is there a link? World J Gastroenterol 20:6374–6385
58. Ryan KJ, Ray CG (2004) Sherris medical microbiology, 4th edn. McGraw Hill, pp 322–324
59. Bien J, Palagani V, Bozko P (2013) The intestinal microbiota dysbiosis and Clostridium difficile infection: is there a relationship with inflammatory bowel disease? Therap Adv Gastroenterol 6:53–68
60. Navaneethan U, Venkatesh PG, Shen B (2010) Clostridium difficile infection and inflammatory bowel disease: understanding the evolving relationship. World J Gastroenterol 16: 4892–4904
61. Nitzan O, Elias M, Chazan B, Raz R, Saliba W (2013) Clostridium difficile and inflammatory bowel disease: role in pathogenesis and implications in treatment. World J Gastroenterol 19:7577–7585

62. Feeney M, Ciegg A, Winwood P, Snook J (1997) A case-control study of measles vaccination and inflammatory bowel disease. Lancet 350:764–766 The East Dorset Gastroenterology Group
63. D'Souza Y, Dionne S, Seidman EG, Bitton A, Ward BJ (2007) No evidence of persisting measles virus in the intestinal tissues of patients with inflammatory bowel disease. Gut 56:886–888
64. Linton MS, Kroeker K, Fedorak D, Dieleman L, Fedorak RN (2013) Prevalence of Epstein-Barr virus in a population of patients with inflammatory bowel disease: a prospective cohort study. Aliment Pharmacol Ther 38:1248–1254
65. Iizuka M, Saito H, Yukawa M, Itou H, Shirasaka T, Chiba M, Fukushima T, Watanabe S (2001) No evidence of persistent mumps virus infection in inflammatory bowel disease. Gut 48:637–641
66. Sipponen T, Turunen U, Lautenschlager I, Nieminen U, Arola J, Halme L (2011) Human herpesvirus 6 and cytomegalovirus in ileocolonic mucosa in inflammatory bowel disease. Scand J Gastroenterol 46:1324–1333

Chapter 7
Clinical Features

Natalia Fabisiak

This chapter focuses on their localization, symptomatology and course of the disease. Additionally, the most frequent extraintestinal manifestations and complications associated with inflammatory bowel diseases are discussed in the section.

7.1 Localization

Crohn's disease (CD) can affect all parts of the gastrointestinal tract, from mouth to anus. The upper part of the alimentary tract is occupied quite rarely—the lesions are observed only in 0.5–13 % of CD patients. These lesions are mostly accompanied by the inflammation in the ileum or large intestine. However, in some cases they can appear exclusively in oral cavity, esophagus, stomach or duodenum.

Frequency of oral manifestations is estimated at 5–20 % cases in adults and 40–80 % in children with CD. The lesions in oral cavity can precede occurrence of a full-blown disease.

Esophageal lesions in CD are usually related with manifestations in ileum and/or colon and exist especially at a younger age. Usually, the distal part of esophagus is involved. Different studies estimate the frequency of these lesions at 7–43 % in children and at 0.2–11 % in adults with CD.

Crohn's disease may affect the stomach and duodenum. Gastroduodenal manifestation is present in 0.5–4.5 % of all CD patients. Gastroduonenal CD the most frequently occupies the antrum of the stomach and the second part of duodenum.

N. Fabisiak (✉)
Department of Biochemistry, Faculty of Medicine, Medical University of Lodz, Mazowiecka 6/8, 92-215 Lodz, Poland
e-mail: nat.fabisiak@gmail.com

J. Fichna (ed.), *Introduction to gastrointestinal diseases Vol. 1*,
DOI 10.1007/978-3-319-49016-8_7

The usual localization of Crohn's disease is the distal part of the ileum. Ileocecal region is occupied in 40–50 % of patients and that is why the disease was called '*ileitis terminalis*' in the past. In 25–30 % of cases CD is limited to the small intestine and in 30–40 % of patients the inflammation is localized simultaneously in both: small and large intestine. The isolated occupation of colon is observed in 20 % of cases.

Anal lesions are characteristic for CD and occur in 35–45 % of patients. Fissures, fistulae or abscesses can precede lesions in intestine or occur concurrently. Lesions in the anal region are more often observed in patients with occupied large bowel, or small and large intestine than in patients with disease solely in the small intestine.

Three forms of ulcerative colitis (UC) are distinguished with regard to localization of the disease: (1) proctitis, when only the rectum is involved, (2) left-side colitis, when lesions are localized distally to the splenic flexure and (3) extensive colitis, when lesions extend proximally to the splenic flexure. Pancolitis, belonging to the extensive colitis, is inflammation of the entire colon.

7.2 Signs and Symptoms

Symptoms presenting in CD and UC show fundamental differences in respect of different localization of lesions in the bowel.

The most common symptom presenting in patients with colonic CD is chronic diarrhea, which lasts more than 6 weeks. The majority of patients complain of persistent abdominal pain or cramps and weight loss. Blood and/or mucus in stool occur in almost half of the patients. Palpable tumor localized in the right lower area of abdomen is present in about one third of the patients. When the disease occupies the upper part of the digestive tract, patients may complain about pain in oral cavity, swallowing difficulties (dysphagia) or pain (odynophagia), epigastric pain, nausea and vomiting. Pain in the region of the anus may indicate inflammation and the formation of fistulae and abscesses.

Chronic, bloody diarrhea, sometimes with passage of mucopurulent exudates, is the primary symptom of patients with UC. Visible blood in the stool occurs in more than 90 % of patients. In the active phase of disease patient may pass up to twenty stools per day. Crampy pain in the lower left area of abdomen, relieved after defecation is reported by patients. Rectal urgency, tenesmus and nocturnal defecation is often described by patients. If lesion are bound only to the terminal part of the colon, especially to the rectum, different symptoms occur. Instead of diarrhea, patients may suffer from constipation, and rectal bleeding may be the only symptom.

Similar nonspecific general manifestation of both diseases are fatigue, tiredness, malaise, anorexia (loss of appetite) or fever.

Table 7.1 Differences between Crohn's disease and ulcerative colitis on the basis of signs and symptoms

Signs and symptoms	Crohn's disease	Ulcerative colitis
Abdominal pain	Often, severe	Less increased
Hemorrhage	Seldom	Very common
Palpable tumor in abdomen	Common	Absence
Fistulae	Common	Seldom
Stenosic lesions	Common	Seldom
Perianal changes	Common	Seldom
Rectum occupation	Seldom	Very common
Distribution of lesions	Continuous	Discontinuous
Toxic megacolon	Seldom	Common
Perforation	Seldom	Common
Pseudopolyps	Quite common	Seldom

Main differences between CD and UC involving signs and symptoms are summarized in Table 7.1. Information about a thorough medical history and examination is shown in Boxes 7.1 and 7.2.

Box 7.1 A full medical history of IBD should include answers to the following questions:

- When did the disease begin? When were the first symptoms?
- Are there recurrent episodes of rectal bleeding or bloody diarrhea?
- Does abdominal pain occur?
- Have you had any problems with stools (tenesmus/incontinence/nocturnal diarrhea)?
- Have you travelled recently?
- Have you got food intolerance?
- Does anyone in your family suffer from inflammatory bowel diseases?
- Have you ever had appendectomy?
- Do you take any medicines, especially antibiotics or non-steroidal anti-inflammatory drugs?
- Do you smoke?
- Have you had any contact with enteric infectious illness recently?
- Have you ever had any problems with your skin, eyes, joints?
- Have you ever had any changes in your mouth or anal area?

Box 7.2 A general psychical examination should include

- general well-being
- body weight and height
- body temperature
- pulse rate
- blood pressure
- abdominal examination (attention for tenderness, distension and palpable masses)
- digital rectal examination
- perianal and oral inspection
- check for eye/skin/joint involvements

7.3 The Course of Disease

Crohn's disease is a chronic, longstanding condition with alternating periods of remission and exacerbation as characteristic features. Longstanding persistent remission after first episode of the disease occurs only in 10–20 % of patients. Progression of the disease proceeds to the fibrosis, formation of stenoses and fistulae. A risk of fistulae formation is estimated at 20–40 % of patients during the overall duration of CD. Young age during onset of the disease, presence of perianal changes and an early beginning of aggressive treatment are negative prognostic factors.

Ulcerative colitis proceeds with periods of exacerbation and remissions. However, in 5 % of patients permanent exacerbation without remission occurs. Another 5 % of patients have one episode of acute symptoms with longstanding period of remission. Remission is defined as a complete reversal of symptoms and lack of changes in endoscopy. In clinical practice, remission is characterized by a reduction in a number of stools without blood and rectal urgency (3 stool per day or less). Presence of bleeding, increased number of stools and intensity change in endoscopic activity indicate relapse of UC.

7.4 Extraintestinal Manifestations

The frequency of extraintestinal manifestations (EIMs) in patients with IBD ranges from 6 to 47 %; more than one EIM is observed in 25 % of patients. Inflammatory involvement of joints, liver, skin and eyes are considered primary manifestations, with two major groups distinguished: (1) extraintestinal immune-related manifestations in IBD, which are associated with intestinal inflammatory activity e.g. arthritis, erythema nodosum, pyoderma gangrenosum, aphthous stomatitis, iritis, uveitis and (2) autoimmune disorders associated with IBD but not correlating with disease activity e.g. insulin-dependent diabetes mellitus, pancreatitis, primary biliary cirrhosis, primary sclerosis cholangitis, Raynaud phenomen, thyroid autoimmune disease and others.

7.4.1 Musculoskeletal Involvement in IBD

Joint manifestations are the most common extraintestinal manifestations in patients with IBD. The inflammatory involvement of joints occurs in 7–25 % of cases and affects equally both: males and females. Arthritis occurs more frequently in patients with colonic disease than small-bowel disease. Peripheral or axial articular involvement can precede, be simultaneous or begin afterward the diagnosis of IBD. Peripheral arthritis is observed in 5–10 % of patients with UC and 10–20 % in cases of CD. It is associated with the skin, mouth and ocular manifestations. Two types of peripheral arthritis are known: (1) pauciarticular arthritis—involves less than five large joints and is strongly related to IBD activity; acute and self-limiting swelling occur and persist for 5–10 weeks; (2) polyarticular arthritis - affects symmetrical five or more small joints and is not associated with the disease activity. Arthritis may last month or years.

Axial arthropathies occur less frequently than peripheral articular involvement in IBD patients and more often affect males than females. Axial arthropathies are not related to intestinal IBD activity and can be grouped into ankylosing spondylitis and sacroiliitis. The prevalence of ankylosis spondylitis in patients with IBD occurs in around 5–10 % of patients. Severe onset of back pain at a younger age is often characterized. Patients may complain of morning stiffness or pain exacerbation by periods of rest. Prevalence of sacroiliitis is observed in up to 25 % of cases. Symptoms are usually absent and the disease is diagnosed radiographically.

7.4.2 Hepatobiliary Manifestation

Hepatobiliary EIMs are common in IBD patients. The most frequent hepatobiliary disease is primary sclerosing cholangitis (PSC). It is a chronic inflammatory disorder of the biliary tree, whose etiology remains unknown. Inflammation, stricture and fibrosis of intra- and extrahepatic bile ducts are characterized by PSC. PSC occurs in 5 % of patients with UC and 2 % of patients with CD. However, 75 % patients with PSC also suffer from ulcerative colitis and 5–10 % of PSC patients have Crohn's disease. The disease more frequently occurs in male than female, especially at the age 30–59. Patients with PSC experience a few suggestive symptoms, e.g. fatigue, pruritus, jaundice and abdominal discomfort; however, 15–70 % of patients are asymptomatic.

7.4.3 Dermatologic Involvements

Erythrema nodosum, pyoderma gangrenousum and oral ulceration are the most common skin manifestations in IBD, usually related to its activity. Erythrema nodosum affects up to 15 % of patients with CD and 10 % of patients with UC. The disease more frequently occurs in women than in men. It is commonly related with the involvement of eye and joint, isolated colonic manifestation and pyoderma gangrenosum. Patients usually present risen, tender, red or violet inflammatory subcutaneous nodules of typically 1–5 cm in diameter. Erythrema nodosum is usually localized on the anterior exterior surface of the lower extremities. It may rarely occur on the face and trunk.

Pyoderma gangrenosum is more severe, very debilitating and fortunately much rarer EIM, appearing more frequently in UC than CD. This chronic skin disorder occurs in about 1–2 % of IBD patients. Conversely, up to 50 % of patients with pyoderma gangrenosum suffer from IBD. Pyoderma gangrenosum is more common in female than male. It is associated with a familial history of UC, initial pancolitis, black African origin, permanent stoma, eye manifestations and erythema nodosum. Lesion usually begins as an erythromatous papule, pustule or nodule evolving quickly into ulcer with irregular, violet borders. The ulcers can be solitary or multiple, unilateral or bilateral and can occupy from several centimeters to an entire limb. Although pyoderma gangrenosum is localized most commonly in the legs, ulcers can appear on any part of the body. New lesions of pyoderma gangrenosum can develop after any type of trauma (pathergy phenomenon).

Prevalence of oral aphthous ulcers (aphthous stomatitis) is at least 10 % of patients with UC and 20–30 % of patients with CD. Ulcerations resolve quickly, when remission is achieved. Aphthous lesions typically occur on the labial and buccal mucosa but may also be located on the tongue and oropharynx. One of the more common lesions in CD in oral cavity are Sutton's aphthous stomatitis. These are circular or oval ulcerations of a bigger size (1–2 cm of diameter), which can occur in every area of oral mucous membrane, most frequently in cheek and velum or in vicinity of the small salivary glands. Routinely they are single, but they can appear in groups. The ulcerations are healed during a few weeks and often recur. They are very painful and hinder eating and drinking.

7.4.4 Ocular Manifestations

Involvement of the eyes occur in 0.3–5 % of IBD patients, more frequently in patients with CD than UC and are often presented with other extraintestinal manifestations, especially peripheral arthritis and erythema nodosum. Three main types of ocular manifestations are distinguished: episcleritis, scleritis and uveitis. Episcleritis is the most common ocular manifestation, defined as painless hyperemia of conjunctiva and sclera. It often parallels intestinal activity. Acute hyperemia, irritation, burning and tenderness to palpation are characteristic symptoms of the disease.

Scleritis is suspected when above symptoms occur with impairment of vision.

Inflammation of the middle chamber of the eyes is called uveitis. Uveitis is less common and occurs independently of disease activity, more frequently in women than man. It is characterized by the ocular pain, visual blurring, photophobia and headache. Uveitis may precede the diagnosis of IBD.

7.5 Extraintestinal Complications

7.5.1 Anemia

About two-third of patients with IBD have anemia, which significantly impairs the quality of life. Iron deficiency anemia and anemia of chronic disease are the most common types of anemia and often the two types overlap. Iron deficiency anemia is the most frequent anemia occurring in IBD which prevalence ranges from 36 to 76 % of patients. The latter type of anemia appears as a result of immune system activation or changes in iron metabolism occurring in patients with any chronic process of active inflammation.

7.5.2 Thromboembolic Events

Patients with IBD have increased risk of developing thromboembolic complications, which are a major cause of morbidity and mortality in IBD. In patients with IBD thrombotic accidents as deep vein thrombosis or pulmonary thromboembolism occur in earlier age than in patients without IBD. Episodes of thrombosis are more frequent in active or complicated IBD and occur mainly in veins.

7.5.3 Osteopathy and Osteoporosis

Inflammatory bowel diseases are related to an increased risk of developing osteopenia and osteoporosis. The prevalence of osteoporosis ranges from 2 to 30 %. Osteopenia occurs in up to 50 % of patients. Osteoporosis in IBD is a multifactorial process. The use of corticosteroids, malabsorption of vitamin D and calcium, low body mass index and the grade of disease activity are important pathogenic factors in IBD.

Chapter 8
Diagnostic Criteria in IBD (with Comments)

Adam Fabisiak

Abstract The reader should already notice the fact that symptoms accompanying IBD are not specific and could be attributed to almost any disease of the lower gastrointestinal tract. Of course, an advanced gastroenterologist or any physician experienced in recognizing IBD would come up with the diagnosis only by the symptoms. Thus, the first major diagnostic tool is a clinical examination, which consists of a thorough medical history and a physical examination. However, to achieve the diagnosis, the process of recognizing the disease should be filled with proper additional studies. In this chapter, tools for setting an appropriate diagnosis, main diagnostic criteria of IBD and diseases from which to distinguish the IBD are discussed. We mostly relied on guidelines by the European Crohn's and Colitis Organisation (ECCO), which gathers the greatest specialists in the field of IBD from around the Europe.

Keywords Diagnosis · Endoscopy · Radiology · Classification

Abbreviation List

IBD	Inflammatory bowel disease
ECCO	European Crohn's and Colitis Organisation
CRP	C-reactive protein
ESR	Erythrocyte sedimentation rate
IBS	Irritable bowel syndrome
ANCA	Anti-neutrophil cytoplasmic antibodies
ASCA	Anti-Saccharomyces cerevisiae antibodies
Anti-OmpC	Antibodies to *Escherichia coli* outer membrane porin C
Anti-CBir1	Antibodies to bacterial flagellin
MRI	Magnetic resonance imaging
US	Ultrasonography
CT	Computed tomography

A. Fabisiak (✉)
Department of Biochemistry, Faculty of Medicine, Medical University of Lodz, Mazowiecka 6/8, 92-215 Lodz, Poland
e-mail: adfabisiak@gmail.com

J. Fichna (ed.), *Introduction to gastrointestinal diseases Vol. 1*,
DOI 10.1007/978-3-319-49016-8_8

SBCE	Small bowel capsule endoscopy
DBE	Double-balloon enteroscopy
CDAI	Crohn's Disease Activity Index

8.1 Tools Used to Diagnose IBD

8.1.1 Laboratory Tests

Initial laboratory tests should be an introduction to the process of diagnosis rather than the main diagnostic tool. In fact, there is no single and perfect serological marker which could—without a doubt—recognize IBD. Nevertheless, some of the basic testing could be performed already in a general practitioner's office and thus facilitate further care of the patient.

First of all, the full blood count should be ran with C-reactive protein (CRP) and erythrocyte sedimentation rate (ESR). Anemia and thrombocytosis are the most common changes seen in full blood count in patients with IBD. Anemia in IBD is pathophysiologically multifactorial and occurs as an overlap of iron-deficiency anemia and anemia of chronic diseases (it could be either a symptom or a complication). Thrombocytosis appears as a result of increased concentration of thrombopoietic stimulators, i.e. acute phase proteins (e.g. thrombopoietin).

Markers evaluating the inflammatory response include CRP and ESR. The former is significantly more important in assessing disease activity as it correlates better with clinical indices and the response in CRP expression to acute inflammation is more rapid than the elevation of ESR.

The best surrogate marker available is calprotectin, a protein which binds calcium in granulocytic cytoplasm. Calprotectin is measured in stool, with sensitivity and specificity in adults estimated at around 93 and 96 %, respectively. Fecal calprotectin is effective in distinguishing IBD from functional disorders such as irritable bowel syndrome (IBS), since no matter what the cause is, the intestinal inflammation will always be accompanied by infiltration of granulocytes (neutrophils). It is also a great tool to avoid unnecessary colonoscopies—negative results of fecal calprotectin most likely rule out IBD. Despite its utility, fecal calprotectin measurement is still a privilege in some of the centers and is not performed routinely in every patient. Therefore, alternatively, stool specimens should be cultured for common intestinal pathogens, such as *Clostridium difficile*, *Yersinia pestis*, *Campylobacter* spp. etc.

There are no viable serologic tests which could help in diagnosis. Several serological markers have been proposed as a potential aid in diagnosing or differentiating IBD, such us:

- Anti-neutrophil cytoplasmic antibodies (ANCA) found in patients with vasculitis
- Anti-Saccharomyces cerevisiae antibodies (ASCA)
- Antibodies to *Escherichia coli* outer membrane porin C (Anti-OmpC)
- Antibodies to bacterial flagellin (anti-CBir1),

but eventually they failed due to their low sensitivity. Therefore to date, there are no recommendations in using any serological markers while diagnosing the patient towards IBD.

8.1.2 Endoscopy

The core procedure of both, diagnosis and management of IBD lays in endoscopic methods. Full colonoscopy with the assessment of terminal part of the ileum should be performed. During the procedure, biopsies from the ileum and each segment of the colon are obtained. In the acute phase of the disease, with massive inflammatory state of the bowel, the mucus is thinner and prone to perforation and colonoscopy is generally not recommended then. In this case, colonoscopy is postponed and should be replaced by the sigmoidoscopy. Nevertheless, the macroscopic and microscopic evaluation of the colon is the key to diagnosis.

The endoscopist assesses the following features:

- vascular pattern
- granularity
- friability of the mucosa
- ulceration
- presence of complications.

Detection of discontinued and asymmetrical lesions with cobblestone appearance in the ileum and the colon, the rectum sparing and perianal changes are in favour of CD. Conversely, the UC is characterized by continuous and symmetrical inflammation with rectum involvement. Figure 8.1 shows the representative findings which could be revealed during endoscopy and Table 8.1 summarizes these findings in regard to disease. Please bear in mind that the listed features are flagships for UC and CD, but the clinical image is not always that consistent and may vary in patients.

Additionally, the pathologist evaluates the samples acquired during the endoscopy. The histopathologic hallmark of CD are granulomas, which are not associated with the intestinal crypt injury and the transmural manner of inflammation (whole intestinal wall is affected). In specimens with UC, inflammation is usually limited to mucosa and appears as a widespread crypt distortion and crypt abscesses. Of note, none of these features has to be present in the early stage of the disease. The most prevalent microscoping abnormality seen within two weeks after the occurrence of the symptoms is basal plasmocytosis.

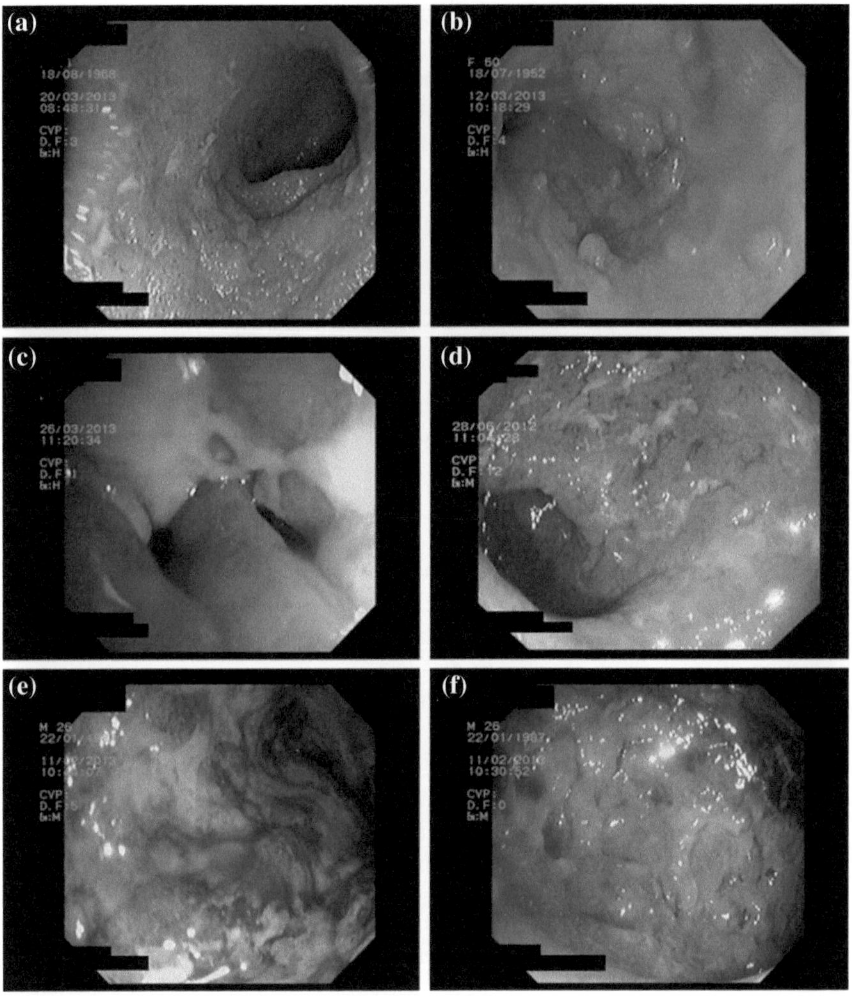

Fig. 8.1 Endoscopic images of pathological changes in the gastrointestinal tract in Crohn's disease (**a–c**) and ulcerative colitis (**d–f**). Crohn's disease: **a** Edematous mucous membrane with slight ulceration covered by fibrin in distal part of small intestine. **b** Pseudopolyps in sigmoid colon. **c** Stenosis of colon at a level of splenic flexure with extensive ulceration covered by fibrin. Ulcerative colitis: **d** Hemorrhagic stadium in sigmoid colon—edematous mucous membrane, redness and friability with flat erosions covered by fibrin. **e** Ulcerative stadium—flat ulceration covered by fibrin. **f** Polypoid stadium—several deep ulcerations covered by fibrin and pseudopolyps

When CD is suspected, the upper GI endoscopy (also called "gastroscopy") is recommended to confirm or exclude the involvement of the upper GI. Importantly, the procedure is of key significance in patients with symptoms suggesting any pathology in the upper GI tract and/or in unclassified colitis. The symptoms are as

Table 8.1 Endoscopic findings in patients with IBD

	Crohn's disease	Ulcerative colitis
Pattern of inflammation	Discontinous	Continous
Involvement of rectum	−	+
Involvement of ileum	+	−
Perianal changes	+	−
Fistulas	+	−
Perforations	−	+
Stenoses	+	−
Mucosal pseudopolyps	+	−

follows: heartburn, upper abdominal pain or discomfort, nausea, and belching. Biopsies should be taken from the duodenum and any suspicious lesions.

8.1.3 *Visualization of the Small Bowel*

In up to one-third of patients, the disease is strictly localized to the small bowel and in around 15 % of patients penetrating lesions develop. A detailed view of small bowel is far more challenging than the large bowel or stomach, even though the past decades brought a significant number or techniques with satisfying accuracy in recognizing lesions in the ileum and the jejunum. Currently, radiological and—to a lesser extent—endoscopic methods constitute a group of tests used in the imaging of small intestine. Following procedures are particularly useful in patients with a suspicion of CD or with unclear image during ileocolonoscopy.

Due to its complexity, visualization of the small bowel requires a specialist with good expertise, but there is still a place for a general practitioner, who can briefly explain the procedure to the patient and meet their any other demands. Nowadays, more and more patients demand from their practitioner a referral for particular procedures (such as magnetic resonance imaging (MRI) or capsule endoscopy). Overall it is a good sign, as these patients are often more engaged in the treatment process. However, the physician should clarify the needs of the patient and whether there is a need to perform particular procedures. Preferably, less invasive tests should be discussed in detail with the patient.

8.1.3.1 Ultrasonography

The most universal and cost-effective procedure is ultrasonography (US). At a first glance, US could be regarded as inefficient due to high interobserver variability and difficulties in viewing deeply situated loops. As a matter of fact, the utility of US in IBD has been proven in both, UC and CD. A recent meta-analysis which sought the diagnostic accuracy of US in detecting CD, showed that the sensitivity and specificity range from 75–94 % and 67–100 % in included studies, respectively (1). Such wide ranges resulted from a discrepancy in deciding on cut-off value of the bowel thickness by the authors of the studies. After statistical analysis of the data, sensitivity and specificity of 88 and 93 % were obtained, respectively, for a threshold of bowel thickness greater than 3 mm; when threshold greater than 4 mm was used, sensitivity and specificity of 75 and 97 %, respectively were achieved. What is more, US can detect colonic or small bowel inflammation with a sensitivity of 80–90 %. Despite the lack of ability to discriminate specific causes of inflammation, US could be regarded as an initial testing because of its noninvasiveness and low cost. Additionally, a recent prospective study reports about the usefulness of US in assessing the response of severe UC to therapy and tendency to accurately predict the course of the disease (2).

8.1.3.2 Computed Tomography and Magnetic Resonance Imaging

Computed tomography (CT) and MRI are regarded as a standard for visualizing the small intestine. Both techniques are able to reveal the intestinal inflammation based on a thickened wall and an increased contrast enhancement. Both procedures also provide the assessment of the extraluminal manifestations and complications such as: abscesses, fistulae, strictures.

Noteworthy, a meta-analysis of thirty-three studies reported no significant differences in accurately diagnosing IBD between US, CT, MR and scintigraphy; hence, CT and MRI both have comparable sensitivity and specificity. Regardless of that fact, MRI should always be considered if there are contraindications for ionizing imaging or the tests will be performed repeatedly.

Both aforementioned examinations require the use of enteral contrast for an appropriate distention of the intestinal lumen. The contrast can be either ingested by the patient (enterography) or can be provided via nasojejunal tube placed past the duodenojejunal flexure under the fluoroscopic control (enteroclysis). The latter is less convenient for the patient and increases exposure to radiation. This and the fact that enterography grants adequate distention makes enteroclysis reserved for cases in which satisfactorily distention cannot be achieved by oral ingestion of the contrast.

8.1.3.3 Endoscopy of the Small Bowel

Endoscopic techniques used to recognize the lesions in small bowel consist of small bowel capsule endoscopy (SBCE) and double-balloon enteroscopy (DBE). SBCE has a higher sensitivity in diagnosing lesions in the small bowel than the radiological methods, especially the superficial ones. However, this method has to be preceded by the exclusion of strictures which could serve as a blockade in the passage of capsule. Hence, the small bowel imaging should be performed before the SBCE due to high prevalence of partial obstructions in CD. Thus, capsule endoscopy is usually reserved for patients with a high suspicion of CD with unclear signs found during other technics (ileocolonoscopy, enteroclysis, CT, MRI).

Double-balloon endoscopy is also a rather young method, invented in the year 2001. Firstly, the patient is sedated for the procedure. Enteroscope is advanced into the duodenum in a conventional manner as in a gastroduodenoscopy. Then, one of the balloons at the tip of the endoscope is inflated and the overtube is being pressed distally. Endoscopist inflates the proximal balloon, deflates the distal one and pushes the endoscope further viewing the segment of the bowel. Finally, the distal balloon is inflated and the endoscope is pulled shortening the intestine. By repeating these three steps it is possible to view the entire small bowel in most of the patients. What is more, the procedure provides an opportunity to take biopsies and perform therapeutic interventions. The key disadvantage of DBE is the time required; it varies from 75 min up to three hours. Main complications are: abdominal pain (20 %), acute pancreatitis (up to 2 %) or asymptomatic hyperamylasemia/hyperlipasemia (20 %). DBE could be also done through the colon and ileum, visualizing the small bowel retrogradely.

Both of the above mentioned procedures are highly specialist and yet of restricted availability. Thus, they are limited to particular cases; in these situations SBCE is preferred to DBE.

8.1.4 Classification

The treatment in IBD has gone through different approaches. Nowadays, guidelines strongly encourage an individual and tailored clinical management. It is recommended to not apply the treatment until specific clinical features of the individual patient's disease are identified. Obviously, the recommendation does not embrace severe disease, in that case urgent surgery is necessary. Nonetheless, precise analysis of the following characteristics is suggested: activity, extent and behavior of the disease. Although a plethora of activity indices exists, the guidelines encourage to use certain scales to measure the disease activity. Favoured indices are the Crohn's Disease Activity Index (CDAI) for CD and either Truelove and Witts' criteria or modified Mayo score for UC. Of note, every disease activity index suffers from disadvantages and none is fully validated. Thus, they are usually not applied in clinical practice and a simpler assessment is conducted. In UC, a combination of clinical features, laboratory results, imaging and endoscopic findings is preferred. In CD, objective signs of active disease such as markers of inflammation and endoscopic activity are needed. Albeit the activity indices are generally cumbersome, they can appear useful for a practitioner in complicated cases.

Additionally, classifications are used to address the time of onset, extent (described further in the chapter) and behaviour of the disease. Montreal classification is advocated to be used worldwide (Table 8.2). The age on diagnosis is of significance: younger patients tend to have more severe disease with higher requirements

Table 8.2 Montreal classification for inflammatory bowel diseases with Paris modification dedicated to young patients

	Montreal	Paris
Age of diagnosis	A_1, below 17 years A_2, between 17 and 40 years A_3, above 40 years	A_{1a}, between 0 and 10 years A_{1b}, between 10 and 17 years A_2, between 17 and 40 years A_3, above 40 years
Location	L_1, ileal L_2, colonic L_3, ileocolonic L_4, separated upper GI tract	L_1, disease in 1/3 of distal ileum L_2, colonic L_3, ileocolonic L_{4a}, upper GI tract disease proximal to ligament of Treitz[a] L_{4b}, upper GI tract disease distal to ligament of Treitz[a]
Behavior	B_1, non-stricturing, non-penetrating B_2, stricturing B_3, penetrating p, perianal disease modifier	B_1, non-stricturing, non-penetrating B_2, stricturing B_3, penetrating B_2B_3, penetrating and stricturing disease p, perianal disease modifier
Growth retardation		G_0, no growth retardation G_1, growth retardation observed

[a]L_4 modifier can be added to L_1, L_2, L_3 if co-exists with upper GI tract disease
Paris modification was created to fully display the juvenile IBD phenotype

for immunomodulators than patients with later onset. Moreover, these patients are at greater risk of colorectal cancer in future due to longer duration of the disease. Allocation to a specific phenotype at the onset facilitates prediction of the disease pattern. All of this features reflect initial therapy, from which the patient would benefit the most.

8.1.5 *Differential Diagnosis*

On the first admission, especially in patients with milder course of disease, differential diagnosis should be performed. There are no pathognomonic traits which definitely indicate IBD. This means that alternative diagnoses ought to be excluded, adding to the whole diagnostic process.

Alternative diagnoses include the diseases which occupy the gastrointestinal tract. Majority of those are infectious diseases caused by a variety of pathogens, among others: *Yersinia enterocolitica*, *Campylobacter*, *Sallmonella*, *Cytomegalovirus*, *Entamoeba histolytica*. Gastrointestinal tuberculosis shares the resemblance with CD and has to be taken into consideration in patients from endemic countries or travelling from these regions. Other diseases include: pseudomembranous colitis induced by *Clostridium difficile*, vasculitis and drug-induced colitis.

Importantly, specific groups of patients should be addressed. These groups comprise of: patients with either a very early (<6 years old) or late onset, as well as immunosuppressed patients. The former group often presents an atypical phenotype of IBD, leading to the diagnosis of unclassified IBD in up to 20 %. Moreover, studies show a rather high prevalence of monogenic diseases such as immunological defects in younger patients with IBD-like phenotype (especially the case of resistant to conventional therapy and refractory IBD should raise a suspicion of a rare monogenic disease).

Conversely, in elder patients ischemic colitis and diverticulitis must be excluded. Ischemia-induced colitis usually appears as an acute state with abdominal pain and rectal bleeding. In certain cases it may occur as a chronic process, for instance when the vessel is not fully obstructed. In ischemic colitis pathologies resembling those seen in IBD could be developed such as pseudopolyps and strictures. Colitis associated with diverticular disease may also mimic IBD, especially CD. Symptoms are non-specific: abdominal pain or discomfort, rectal bleeding, nausea, altered defecation. Normally, the disease is benign but some patients may develop complications such as: pericolonic abscesses, perforations, intestinal blockage and fistulae. In immunosuppressed patients, colitis may present in two ways: due to opportunistic infections or infestations and due to administered drugs (such as methotrexate or mycophenolate mofetil). Diagnosing of the infectious colitis should always be managed in regard to the local epidemiologic situation. Stool samples and blood tests should be taken for various antigens and antibodies titres.

Distinction between CD and UC is of key significance. Completing the whole diagnostic process and a thoughtful consideration of the gathered data lead to correct and/or better treatment outcomes. It is particularly relevant when considering the surgical therapy; up to 90 % of patients with CD who undergo restorative procto-colectomy with ileal-pouch-anal-anastomosis develop complications, while the majority of UC patients with such a treatment do well. However, in some cases it is not possible to tell the difference and patients are diagnosed with indeterminated colitis or misdiagnosed. Hence, patients with uncertain diagnosis should be monitored and eventually their diagnosis should be clarified.

Overall, it is crucial for patient to understand the need of compliance with a physician and his entanglement in disease monitoring.

1. Fraquelli M, Colli A, Casazza G, Paggi S, Colucci A, Massironi S, et al. Role of US in detection of Crohn disease: meta-analysis. Radiology. 2005;236(1):95–101.
2. Parente F, Molteni M, Marino B, Colli A, Ardizzone S, Greco S, et al. Are colonoscopy and bowel ultrasound useful for assessing response to short-term therapy and predicting disease outcome of moderate-to-severe forms of ulcerative colitis?: a prospective study. Am J Gastroenterol. Nature Publishing Group; 2010;105(5):1150–7.

All guidelines which were used during the preparation of the chapter are found at https://www.ecco-ibd.eu/index.php/publications/ecco-guidelines-science/published-ecco-guidelines.html.

Acknowledgments Authors would like to acknowledge the senior doctor Anna Mokrowiecka Ph.D. MD and the Department of Gastroenterology at the Barlicki Hospital, Medical University of Lodz for providing the endoscopic images of patients with IBD.

Chapter 9
Clinical Treatment in IBD

Marcin Włodarczyk and Aleksandra Sobolewska-Włodarczyk

Abstract Nowadays, treatment of IBD is still controversial. The management plan for a patient with Crohn's disease should take into account the activity, site and behavior of disease, and should always be discussed with the patient. When deciding the appropriate treatment strategy for active ulcerative colitis, one should consider the activity, distribution and pattern of disease (relapse frequency, course of disease, response to previous medications, side-effect profile of medication and extra-intestinal manifestations). The age at onset and disease duration may also be important factors. Generally, in both diseases an individual approach to each patient cannot be neglected. The goal of the treatment, especially maintenance therapy, in both UC and CD is to achieve and maintain a steroid-free remission, clinically and endoscopically defined. In this chapter groups of drugs and their guidance for use will be discussed.

Keywords Inflammatory bowel disease · Treatment · Steroids · 5-ASA · Anti-tumor necrosis factor alfa

M. Włodarczyk (✉) · A. Sobolewska-Włodarczyk (✉)
Department of Biochemistry, Faculty of Medicine,
Medical University of Lodz, Mazowiecka 6/8, 92-215 Lodz, Poland
e-mail: dr.mwlodarczyk@gmail.com

A. Sobolewska-Włodarczyk
e-mail: olasobolewska1@poczta.onet.pl

M. Włodarczyk
Department of General and Colorectal Surgery, Faculty of Military Medicine,
Medical University of Lodz, Lodz, Poland

A. Sobolewska-Włodarczyk
Department of Gastroenterology, Faculty of Military Medicine,
Medical University of Lodz, Lodz, Poland

J. Fichna (ed.), *Introduction to gastrointestinal diseases Vol. 1*,
DOI 10.1007/978-3-319-49016-8_9

1. Drugs in IBD treatment

(A) **Corticosteroids**

Corticosteroids are a class of steroid hormones that are produced in the adrenal cortex of vertebrates, as well as their synthetic analogues. Corticosteroids are involved in a wide range of physiological processes, including stress response and immune response, regulation of inflammation, carbohydrate metabolism, and protein catabolism, blood electrolyte levels, and behavior [1–3].

Synthetic pharmaceutical drugs with corticosteroid-like effects are used in a variety of conditions, including IBD.

In the treatment of IBD hydrocortisone, budesonide and methylprednisolone are commonly in use.

Hydrocortisone:
(Hydrocortisone is a name for cortisol when it is used as a medication)

- Stimulates gluconeogenesis (formation of glucose), and activates anti-stress and anti-inflammatory pathways.
- Counteracts insulin; contributes to hyperglycemia, stimulating hepatic gluconeogenesis and inhibiting peripheral utilization of glucose (insulin resistance).
- Reduces bone formation, favoring long-term development of osteoporosis (progressive bone disease).
- Raises free amino acid levels in the serum.
- Acts as a diuretic by increasing water diuresis, glomerular filtration rate, and renal plasma flow from the kidneys, as well as stimulating sodium retention and potassium excretion. It also increases sodium and water absorption and potassium excretion in the intestines [4, 5].

Budenoside:

- Controls the rate of protein synthesis.
- Depresses migration of polymorphonuclear leukocytes and fibroblasts
- Reverses capillary permeability and lysosomal stabilization at the cellular level to prevent or control inflammation.
- Has a potent glucocorticoid activity and weak mineralocorticoid activity [6–8].

Methylprednisolone:
The anti-inflammatory actions of methyloprednizolone are thought to involve phospholipase A2 inhibitory proteins, lipocortins, which control the biosynthesis of potent mediators of inflammation such as prostaglandins and leukotrienes.

Side effects of corticosteroids:

- increased appetite
- acne
- rapid mood swings and mood changes—such as becoming aggressive, irritable and short-tempered with people
- thin skin that bruises easily

- muscle weakness
- delayed wound healing
- a combination of fatty deposits that develop in the face, stretch marks across the body and acne—known as the Cushing's syndrome
- weakening of the bones (osteoporosis)
- diabetes (or they may worsen existing diabetes)
- high blood pressure
- glaucoma and cataracts (eye conditions)
- stomach ulcers—one may be prescribed an additional medication called a proton pump inhibitor (PPI) to reduce this risk
- increased risk of infections, particularly chickenpox, shingles and measles
- reduced growth in children [9, 10].

(B) Derivatives of 5-aminosalicylic acid (5-ASA): mesalazine and sulfasalazine

Mesalazine or 5-aminosalicylic acid is a bowel-specific aminosalicylate drug that acts locally in the gut and has its predominant actions there, thereby having few systemic side effects. As a derivative of salicylic acid, mesalazine is also thought to be an antioxidant that traps free radicals, which are potentially damaging byproducts of metabolism. An active moiety of sulfasalazine, which is metabolized to sulfapyridine and mesalazine.

Sulfasalazine and its metabolite 5-ASA are poorly absorbed from the small intestine; its main mode of action is therefore believed to be inside the intestine. Approximately one third of a dose of sulfasalazine is absorbed from the small intestine. The remaining two thirds pass into the colon where the drug is split by bacteria into 5-ASA and sulfapyridine. Sulfapyridin eis well absorbed from the colon (estimated bioavailability 60 %); 5-ASA is less well absorbed (estimated bioavailability 10–30 %).

Side effects of 5-ASA:

- Diarrhea
- Nausea
- Cramping
- Flatulence
- Headache
- Hypersensitivity reactions (including rash, urticaria *aka* hives, interstitial nephritis and lupus erythematosus-like syndrome)
- Hair loss
- Acute pancreatitis
- Hepatitis
- Nephrotic syndrome
- Blood disorders (including agranulocytosis, aplastic anaemia, leukopenia, neutropenia, thrombocytopenia).

Mesalazine avoids the sulfonamide side effects of sulfasalazine, which contains additional sulfapyridine, but carries additional rare risks of: allergic lung reactions, allergic myocarditis, methaemoglobinaemia [11–14].

(C) Thiopurines: azathioprine, tioguanine, mercaptopurine

The thiopurine drugs are purine antimetabolites widely used in the treatment of the inflammatory disease.

The purine molecule is the framework for two of the four bases that occur in DNA, adenine and guanine. Consequently, blocking the synthesis of purine also hinders DNA synthesis and thus inhibits the proliferation of cells, especially fast-growing cells without a method of nucleotide salvage ("recycling"), such as lymphocytes—T-cells and B-cells.

Azathioprine

The active metabolite of azathioprine, methyl-thioinosine monophosphate is a purine synthesis inhibitor that works by blocking the enzyme amidophosphoribosyltransferase.

Side effects of thiopurine:

- Nausea and vomiting (especially at the beginning of the treatment).
- Hypersensitivity reactions include dizziness, diarrhea, fatigue, and skin rashes.
- Hair loss.
- Bone marrow suppression (anaemia).
- Susceptibility to infection.
- Acute pancreatitis can also occur, especially in patients with Crohn's disease [15–21].

(D) Methotrexate

Methotrexate is an antimetabolite and an antifolate drug. Methotrexate competitively inhibits dihydrofolate reductase (DHFR), an enzyme that participates in the tetrahydrofolate synthesis. Methotrexate inhibits the synthesis of DNA, RNA, thymidylates, and proteins. This drug also takes part in the inhibition of enzymes involved in purine metabolism, leading to accumulation of adenosine, inhibition of T cell activation and suppression of intercellular adhesion molecule expression by T cells, selective down-regulation of B cells, increasing CD95 sensitivity of activated T cells, and inhibition of methyltransferase activity, leading to (de)-activation of enzyme activity relevant to immune system function. Another mechanism of MTX is the inhibition of the binding of interleukin 1-beta to its cell surface receptor.

Side effects of methotrexate:

- Hepatotoxicity (liver damage)
- Ulcerative stomatitis
- Bone marrow suppression—low white blood cell count and thus predisposition to infection

- Nausea, abdominal pain, fatigue
- Fever
- Acute pneumonitis, rarely pulmonary fibrosis
- harmful to fetus (pregnancy category X) [22–24].

(E) Cyclosporine

Cyclosporine binds to the cytosolic protein cyclophilin of lymphocytes, especially T cells. The complex of cyclosporine and cyclophilin inhibits calcineurin, which, under normal circumstances, is responsible for activating the transcription of interleukin 2.

Side effects of cyclosporine.

- Enlargement of the gums
- Convulsions
- Peptic ulcers
- Pancreatitis
- Hypercholesterolemia
- High blood pressure
- Potassium retention possibly leading to hyperkalemia
- Nephrotoxicity and hepatotoxicity [25–27].

(F) Tacrolimus

Tacrolimus is an anti-inflammatory drug inhibiting the production of interleukin-2, a molecule that promotes the development and proliferation of T cells. Tacrolimus is a macrolide calcineurin inhibitor.

Side effects of tacrolimus:

- Infection
- Cardiac damage
- Hypertension
- Blurred vision
- Nephrotoxicity
- Hyperkalemia, hypomagnesemia
- Hyperglycemia, diabetes mellitus
- Lung damage
- Various neuropsychiatric complications [28, 29].

(G) Anti-tumor necrosis factor alfa agents

Anti-tumor necrosis factor alpha agent (anti-TNF-alpha) is a pharmaceutical drug that suppresses the immune system. TNF-alpha is involved in autoimmune and immune-mediated disorders. In IBD pathogenesis, TNF-alpha is one of the most important cytokines.

In IBD treatment Infliximab, Adalimumab and Certolizumab are most commonly used.

Side effects of anti-TNF-alpha:

- Increased risk of certain opportunistic infections
- Tuberculosis
- Fungal infection
- Cancerogenesis.

Infliximab is a chimeric monoclonal antibody biologic drug that works against TNF-alpha.
Adalimumab is a fully human monoclonal antibody.
Certolizumab pegol is a PEGylated Fab' fragment of a humanized TNF inhibitor monoclonal antibody [30–36].

9.1 Treatment of Crohn's Disease

1. Treatment according to site of disease and disease activity

(a) Mildly active localized ileocaecal Crohn's disease

Budesonide 9 mg daily is the preferred treatment. The benefit of mesalazine is limited. Antibiotics cannot be recommended. No treatment is an option for some patients with mild symptoms.

(b) Moderately active localized ileocaecal Crohn's disease

Budesonide 9 mg/day or with systemic corticosteroids are recommended. Antibiotics can be added if septic complications are suspected. Thiopurines, especially azathioprine, or methotrexate in combination with steroids is also an appropriate option. Anti-TNF therapy should be considered as an alternative for patients with objective evidence of active disease, who have previously been steroid-refractory, -dependent, or -intolerant. Risks should be carefully considered and discussed with patients.

(c) Severely active localized ileocecal Crohn's disease

Systemic corticosteroids are the treatment of choice. Anti-TNF therapy (especially infliximab) with or without an immunomodulator is an appropriate option for patients with objective evidence of active disease. For some patients who have infrequently relapsing disease, restarting steroids with an immunomodulator may be appropriate. Surgery is a reasonable alternative for some patients and should also be considered and discussed.

(d) Colonic disease

Active colonic Crohn's Disease may be treated with sulfasalazine if only mildly active or with systemic corticosteroids. Colonic disease should be treated as severe active ileocaecal disease.

(e) Extensive small bowel disease

Extensive small bowel Crohn's disease should be treated with systemic corticosteroids and thiopurines or methotrexate. For patients who have relapsed, anti-TNF therapy with or without azathioprine is an appropriate option if there is objective evidence of moderate or severely active disease. The small intestine inflammation impairs absorption of food nutrients often causing malnutrition. Because of this, adjunctive nutritional support is appropriate. Surgical options should also be considered and discussed at an early stage.

Patients who have clinical features that suggest poor prognosis currently appear to be the most suitable subjects for early introduction of thiopurines, methotrexate and or anti-TNF therapy.

(f) Oesophageal and gastroduodenal disease

CD with this localization may be treated with a proton pump inhibitor (PPI). If necessary, together with systemic corticosteroids and thiopurines or methotrexate. Anti-TNF therapy is an alternative for severe or refractory disease. Dilatation or surgery is appropriate for obstructive symptoms.

2. Medical management of patients in pharmacologically induced remission

(a) First presentation of localised disease

After the first presentation, if remission has been achieved with systemic steroids, a thiopurine or methotrexate should be considered. Nowadays, there is no consistent evidence for efficacy of oral 5-ASA. No maintenance treatment is an option for some patients.

(b) Relapse of localised disease

In case of a relapse, escalation of the maintenance treatment can be considered. Steroids should not be used to maintain remission. Surgery should always be considered as an option in localized disease.

(c) Steroid-dependent Crohn's disease

Patients who are dependent on corticosteroids should be treated with thiopurines or methotrexate with or without anti-TNF therapy (infliximab or adalimumab are prefer), although surgical options should also be considered and discussed.

(d) Relapse while on azathioprine

Patients receiving azathioprine or mercaptopurine who relapse should be evaluated for adherence to therapy and have their dose optimised. Change of their maintenance therapy to methotrexate or anti-TNF therapy should be considered. Surgery should always be considered as an option in localised disease.

(e) Maintenance after induction of remission with anti-TNF therapy

In case of remission achieved with an anti-TNF agent, maintenance with regular anti-TNF therapy should be considered. Azathioprine may be considered in combination with anti-TNF therapy or is an option as monotherapy [37, 38].

9.2 Treatment of Ulcerative Colitis

1. Treatment according to site of disease and disease activity

(a) Proctitis

A mesalazine 1 g suppository once daily is the treatment of choice for mild or moderately active proctitis. Suppositories may deliver drug to the rectum more effectively and are better tolerated than enemas. Combining topical mesalazine with oral mesalazine or topical steroid is more effective than either alone and should be considered for escalation of treatment. Oral mesalazine alone is less effective. Unruly proctitis may require treatment with immunosuppressants and/or anti-TNF therapy.

(b) Left side ulcerative colitis

Mild or moderate severity should initially be treated with an aminosalicylate enema 1 g/day combined with oral mesalazine >2 g/day. Topical therapy with steroids or aminosalicylates alone or as mono-therapy with oral aminosalicylates, is less effective than oral plus topical 5-ASA therapy.

Systemic corticosteroids are appropriate if symptoms of active colitis do not respond to mesalazine. In case of severe left-sided colitis, hospitalization is necessary.

(c) Extensive ulcerative colitis

Mild or moderate extensive ulcerative colitis should initially be treated with oral 5-ASA >2 g/day, which should be combined with topical mesalazine to increase remission rates. Systemic corticosteroids are appropriate if symptoms of active colitis do not respond to mesalazine. Severe extensive colitis is an indication for hospital admission for intensive treatment.

In case of severe UC, systemic corticosteroids are used, especially methylprednisolone or hydrocortisone. The response to intravenous steroids is best assessed objectively around the third day of therapy. Treatment options including colectomy should be discussed with patients with severely active UC not responding to intravenous steroids. Second line therapy with either cyclosporine or anti-TNF alpha (mostly infliximab) or tacrolimus may be appropriate. If there is no improvement within 4–7 days of salvage therapy, colectomy is recommended.

2. Maintenance of remission.

Maintenance treatment is recommended for all patients.

Oral 5-ASA is the first line maintenance treatment in patients responding to 5-ASA or steroids. Rectal 5-ASA is the first line in maintenance in case of proctitis and an alternative in left-sided colitis. A combination of oral and rectal 5-ASA can be used as a second line maintenance treatment.

Minimum effective dose to maintenance remission of oral 5-ASA is 1.2 g per and for rectal treatment 3 g/week. Sulfasalazine is equally or slightly more effective other oral 5-ASA preparations are preferred for toxicity reasons. All the different available preparations of oral 5-ASA are effective.

Azathioprine/mercaptopurine is recommended for patients with mild to moderate disease activity who have experienced early or frequent relapse whilst taking 5-ASA at optimal dose or who are intolerant to 5-ASA. In patients responding to anti-TNF agents, both maintaining remission with azathioprine/mercaptopurine and continuing anti-TNF therapy with or without thiopurines are appropriate. In patients with severe colitis responding to intravenous, azathioprine/mercaptopurine should be considered to maintain remission. However, in patients responding to infliximab continuing infliximab is also appropriate. The prior failure of thiopurines favours maintenance with anti-TNF therapy [37, 38].

9.2.1 Surgery in IBD Management

Surgery plays an important role in the treatment of IBD and an individual approach to each patient should be a priority. For many patients operation is associated with failure of treatment. However, cooperation between patient, gastroenterologist and surgeon is a key to improve patient's quality of life.

In case of UC, colectomy may become causal treatment. In CD, surgery is more complicated because of the need for conserving therapy.

Most importantly, surgical interventions should be performed in specialist referral centers as this eliminates post-operative complications.

9.3 Surgery in UC

Surgical treatment, especially colectomy, is recommended in the acute case when patients do not respond to conservative pharmacotherapy, or if a patient has been taking 20 mg daily or more of prednisolone for more than 6 weeks.

In emergency circumstances, during colectomy for ulcerative colitis the whole rectum and the inferior mesenteric artery should be preserved. This facilitates subsequent pouch surgery. The maximum length of anorectal mucosa between the dentate line and the anastomosis should not exceed 2 cm. When performing a restorative proctocolectomy for ulcerative colitis, a covering loop ileostomy is generally recommended, but it can be avoided in selected cases. The loop ileostomy may be closed in almost all ceases and digestive tract continuity can be restored.

To conclude, restorative proctocolectomy with ileal pouch-anal anastomosis has become the new gold standard in surgical UC treatment. This solution offering patients good quality of life with no stoma after restoration of digestive tract continuity and a preserved anal route of defecation [37, 38].

9.4 Surgery in CD

Nowadays, in complicated CD surgery at an early stage is a valid alternative to medical therapy. The type of surgical procedures depends of symptoms and localization of main lesions. In case of localized ileocaecal CD with obstructive symptoms, with no significant evidence of active inflammation, it should be treated by surgery with ileocaecal resection. Surgical procedures should be considered in all cases of CD with obstructive symptoms. The surgical procedures mainly include resections of obstructed part of bowel. Stricturoplasty is a safe alternative to resection in jejuno-ileal Crohn's disease, with similar short-term and long-term results. Stricturoplasty is a surgical procedure performed in response to scar tissue that has built up in the intestinal wall from inflammatory bowel conditions. When CD is localized in colon and the surgery is necessary, resection only of affected part is preferable (in cases when less than a third of the colon involved). Stricturoplasty in the colon is not recommended. At present, an ileal pouch-anal anastomosis is not recommended in a patient with Crohn's colitis. In some cases the surgical procedure of creation of loop ileostomy is necessary in order to bypass the intestine with inflammatory lesions [37, 38].

Acknowledgments Supported by the National Science Center (2015/17/N/NZ5/00677 to ASW).

References

1. Nussey S, Whitehead S (2001) Endocrinology: an integrated approach. BIOS Scientific Publishers, Oxford
2. Swinburn CR, Wakefield JM, Newman SP, Jones PW (1988) Evidence of prednisolone induced mood change ('steroid euphoria') in patients with chronic obstructive airways disease. Br J Clin Pharmacol 26(6):709–713
3. Blackburn D, Hux J, Mamdani M (2007) Quantification of the risk of corticosteroid-induced diabetes mellitus among the elderly. J Gen Intern Med 17(9):1497–1525
4. Scott E (2011) Cortisol and stress: how to stay healthy. About.com. Retrieved 29 Nov 2011
5. Chyun YS, Kream BE, Raisz LG (1984) Cortisol decreases bone formation by inhibiting periosteal cell proliferation. Endocrinology 114(2):477–480
6. Hoehn K, Marieb EN (2010) Human anatomy & physiology. Benjamin Cummings, San Francisco. ISBN 0-321-60261-7
7. "Budesonide". The American Society of Health-System Pharmacists. Retrieved 2 Dec 2015
8. Silverman J, Otley A (2011) Budesonide in the treatment of inflammatory bowel disease. Expert Rev Clin Immunol 7(4):419–428
9. Pardi DS, Tremaine WJ, Carrasco-Labra A (2016). American Gastroenterological Association institute technical review on the medical management of microscopic colitis. Gastroenterology 150(1):247–274.e11
10. "Medrol (Methylprednisolone—Tablets by Pfizer)". Sfatul Medicului. Retrieved 14 Apr 2013
11. Kruis W, Schreiber I, Theuer D, Brandes JW, Schütz E, Howaldt S, Krakamp B, Hämling J, Mönnikes H, Koop I, Stolte M, Pallant D, Ewald U (2001) Low dose balsalazide (1.5 g twice daily) and mesalazine (0.5 g three times daily) maintained remission of ulcerative colitis but high dose balsalazide (3.0 g twice daily) was superior in preventing relapses. Gut 49(6):783–789

12. Sandborn WJ, Feagan BG, Lichtenstein GR (2007) Medical management of mild to moderate Crohn's disease: evidence-based treatment algorithms for induction and maintenance of remission. Aliment Pharmacol Ther 26(7):987–1003
13. Patient information: Disease modifying antirheumatic drugs (DMARDs) (Beyond the Basics). UpToDate. May 2012. Retrieved 09 Dec 2012
14. Oakley Fiona, Meso Muriel, Iredale John P, Green Karen, Marek Carylyn J, Zhou Xiaoying, May Michael J, Millward-Sadler Harry, Wright Matthew C, Mann Derek A (2005) Inhibition of inhibitor of κB kinases stimulates hepatic stellate cell apoptosis and accelerated recovery from rat liver fibrosis. Gastroenterology 128(1):108–120
15. Evans WE (2004) Pharmacogenetics of thiopurine S-methyltransferase and thiopurine therapy. Ther Drug Monit 26(2):186–191
16. Sahasranaman S, Howard D, Roy S (2008) Clinical pharmacology and pharmacogenetics of thiopurines. Eur J Clin Pharmacol 64(8):753–767
17. Elion G (1989) The purine path to chemotherapy. Science 244(4900):41–47
18. Weersma RK, Peters FTM, Oostenbrug LE, van den Berg AP, van Haastert M, Ploeg RJ, Posthumus MD, Homan van der Heide JJ, Jansen PLM, van Dullemen HM (2004) Increased incidence of azathioprine-induced pancreatitis in Crohn's disease compared with other diseases. Aliment Pharmacol Ther 20(8):843–850
19. Maltzman JS, Koretzky GA (2003) Azathioprine: old drug, new actions. J Clin Invest 111 (8):1122–1124
20. Rossi S (ed) (2013) Australian medicines handbook, 2013th edn. The Australian Medicines Handbook Unit Trust, Adelaide
21. Joint Formulary Committee (2013) British national formulary (BNF), 65th edn. Pharmaceutical Press, London, UK
22. Rajagopalan PTR, Zhang Z, McCourt L, Dwyer M, Benkovic SJ, Hammes GG (2002). Interaction of dihydrofolate reductase with methotrexate: ensemble and single-molecule kinetics. Proc Natl Acad Sci 99(21):13481–13486
23. Herfarth HH, Long MD, Isaacs KL (2012) Methotrexate: underused and ignored? Dig Dis 30 (Suppl 3):112–118
24. Ganong WF Review of medical physiology, 22nd edition, Lange medical books, chapter 27, p 530
25. Naesens M, Kuypers DR, Sarwal M (2009) Calcineurin inhibitor nephrotoxicity. Clin J Am Soc Nephrol 4(2):481–509
26. Robert N, Wong GW, Wright JM (2010) Effect of cyclosporine on blood pressure. Cochrane Database Syst Rev (1):CD007893
27. Liu J, Farmer J, Lane W, Friedman J, Weissman I, Schreiber S (1991) Calcineurin is a common target of cyclophilin-cyclosporin A and FKBP-FK506 complexes. Cell 66(4):807–815
28. Baumgart DC, MacDonald JK, Feagan BG (2008) Tacrolimus (FK506) for induction of remission in refractory ulcerative colitis. Cochrane Database Syst Rev 16(3):CD007216 (Baumgart DC (ed))
29. Hanifin JM, Paller AS, Eichenfield L, Clark RA, Korman N, Weinstein G, Caro I, Jaracz E, Rico MJ, US Tacrolimus Ointment Study Group (2005) Efficacy and safety of tacrolimus ointment treatment for up to 4 years in patients with atopic dermatitis. J Am Acad Derm 53(2 suppl 2):S186–94
30. Peppel K, Crawford D, Beutler B (1991) A tumor necrosis factor (TNF) receptor-IgG heavy chain chimeric protein as a bivalent antagonist of TNF activity. J Exp Med 174(6):1483–1489
31. Dubinsky MC, Fleshner PP (2003) Treatment of Crohn's disease of inflammatory, stenotic, and fistulizing phenotypes. Curr Treat Options Gastroenterol 6(3):183–200
32. Present DH, Rutgeerts P, Targan S, Hanauer SB, Mayer L, van Hogezand RA, Podolsky DK, Sands BE, Braakman T, DeWoody KL, Schaible TF, van Deventer SJ (1999) Infliximab for the treatment of fistulas in patients with Crohn's disease. N Engl J Med 340(18):1398–1405
33. Brekke OH, Sandlie I (2003) Therapeutic antibodies for human diseases at the dawn of the twenty-first century. Nat Rev Drug Discov 2(1):52–62
34. Podolsky Daniel K (2002) Inflammatory bowel disease. N Engl J Med 347(6):417–429

35. Schreiber S et al (2005) Certolizumab pegol, a humanised anti-TNF pegylated FAb' fragment, is safe and effective in the maintenance of response and remission following induction in active Crohn's disease: a phase 3 study (precise). Gut 54(suppl7):A82
36. Sandborn WJ, Feagan BG, Stoinov S et al (2007) Certolizumab pegol for the treatment of Crohn's disease. N Engl J Med 357(3):228–238
37. Dignass A, Van Assche G, Lindsay JO, Lémann M, Söderholm J, Colombel JF, Danese S, D'Hoore A, Gassull M, Gomollón F, Hommes DW, Michetti P, O'Morain C, Öresland T, Windsor A, Stange EF, Travis SPL The second European evidence-based Consensus on the diagnosis and management of Crohn's disease: Current management
38. Dignass A, Lindsay JO, Sturm A, Windsor A, Colombel J-F, Allez M, D'Haens G, D'Hoore A, Mantzaris G, Novacek G, Öresland T, Reinisch W, Sans M, Stange E, Vermeire S, Travis S, Van Assche G Second European evidence-based consensus on the diagnosis and management of ulcerative colitis Part 2: Current management

Chapter 10
IBD Patient's Guide

Hubert Zatorski

Abstract Understanding inflammatory bowel disease (IBD) and its treatment poses a great challenge for both the patient and the doctor. In this chapter, diet and lifestyle, psychological aspects and treatment of IBD are discussed. The impact of diet, smoking and alcohol drinking on IBD is also presented. Due to the fact that patients with IBD more often suffer from depression and anxiety, standard treatments for these conditions are also described. Moreover, differences in bacterial content in gastrointestinal (GI) tract between IBD patients and healthy individuals are characterized. Finally, information about cooperation between the patient and the doctor is given.

Keywords Diet · Lifestyle · Microbiota · Obesity · Psychological aspects · Smoking

Abbreviations

AIEC	Adherent-invasive *E. coli*
CD	Crohn's disease
FODMAP	Fermentable Oligo-, Di-, Monosaccharides and Polyols
IBD	Inflammatory Bowel Disease
NSAIDs	Non-steroidal anti-inflammatory drugs
QoL	Quality of Life
SNRI	Serotonin Norepinephrine Reuptake Inhibitors
SSRIs	Selective Serotonin Reuptake Inhibitors
UC	Ulcerative colitis

H. Zatorski (✉)
Faculty of Medicine, Department of Biochemistry, Medical University of Lodz, Mazowiecka 6/8, 92-215 Lodz, Poland
e-mail: zatorski.h@gmail.com

J. Fichna (ed.), *Introduction to gastrointestinal diseases Vol. 1*,
DOI 10.1007/978-3-319-49016-8_10

10.1 Diet and Lifestyle

10.1.1 Diet

Worldwide, the incidence of CD varies from 0.7 to 14.6 individuals and that of UC from 1.5 to 24.5 individuals per 100,000 inhabitants, depending on the region [1]. High incidence of IBD was reported in North Europe and North America, especially in highly developed countries, such as Canada, Iceland, United Kingdom and Australia [2]. A steady increase has been observed in the Northern developed countries in the 1940s, followed by the Southern countries in the 1960s [3, 4]. Recent studies connected this discrepancy in epidemiological data with differences in diet and lifestyle between Western and other countries. Certain foods have been shown to be either beneficial or detrimental for IBD patients, indicating the need for the development of individualized diets. For instance, high vegetable intake has been associated with decreased risk of UC, whereas greater intake of fiber and fruit - with decreased risk of CD [5]. Moreover, a westernized diet rich in animal fat and protein and low in fiber has been correlated with IBD [5]. Other studies have also shown associations between CD and a high intake of carbohydrate, starch and refined sugar [6]. Despite the high number of studies, there is still some controversy about the exact role of diet in IBD development. Nevertheless, there are indications that diet may modulate disease onset and activity.

High intake of monosaccharides and saturated fats as well as low intake of fiber are connected with increased risk of IBD. Studies performed by Sakamoto et al. and Russel et al. showed that an increased consumption of monosaccharides, such as cola-type drinks, chocolate and other sweets increases the risk of developing both CD and UC [7, 8]. According to a study performed by Reif et al. [9], an increased consumption of animal protein may result in a higher risk of IBD. Another report demonstrated that an increased consumption of trans-fatty acids is associated with greater risk of UC [10].

On the other hand, a fiber-rich diet may be beneficial as fiber exerts protective effect on IBD development. According to Ananthakrishnan et al., diet containing 24.3 g of fiber per day may reduce the risk of CD development by 40 % [11]. Positive effect was noted especially with fiber derived from fruit sources. Nevertheless, this effect was not observed in UC patients [11].

It is indisputable that patient's diet should vary during relapses and remissions. Diet with low fiber intake is recommended for patients with IBD during periods of disease exacerbation, diarrhea and abdominal pain, except patients with UC and rectal involvement who may develop constipation [12]. Furthermore, patients with inadequate nutritional supply may be recommended to remain on exclusive enteral nutrition containing all the necessary nutrients [12]. While diet in patients with exacerbated IBD remains clear, recommendations for patients with remission are equivocal. The explanation of this fact is that official recommendations do not address the subject of nutrition of IBD patients and different research groups presents contradictory results. Nevertheless, the National Clinical Guide Center

Table 10.1 Description of diets and their impact on IBD symptoms

Diet	General description	Results	References
Gluten-free diet	Exclude wheat, rye, barley	Improvement of symptoms	[14]
Mediterranean diet	Reduce meat Increase bread, fish, fruits, vegetables and olive oil	Decreased CRP levels after 6 weeks	[15]
Specific carbohydrate diet	Exclude processed meats and all sugars other than monosaccharides	Decreased Crohn's disease activity (Harvey-Bradshaw Index) at week 12 in 9/10 patients	[16]
FODMAP diet	Exclude many fruits, vegetables and legumes, wheat, rye, milk	Improvement of symptoms	[17]

advises patients in remission to be on a diversified and well-balanced diet, but does not provide readers with a detailed composition [12].

Interestingly, in a recent survey organized by Crohn's and Colitis Foundation of America individuals pointed out the foods they believed to ameliorate or exacerbate disease activity. Whereas yogurt and rice were more often reported to improve symptoms, fruits, vegetables, high fiber foods, red meat, fried food and alcohol were more frequently reported to worsen symptoms [13]. Many different types of diets have been proposed in the treatment of IBD, such as gluten-free diet, Mediterranean diet as well as FODMAP (Fermentable Oligo-, Di-, Monosaccharides and Polyols) diet and specific carbohydrate diet, but there is no sufficient data to unambiguously recommend them. Brief description of mentioned diets can be found in Table 10.1.

10.1.2 Lifestyle

10.1.2.1 Smoking

Nowadays, it is unquestionable that smoking negatively affects our health. Smoking is a leading cause of cancer (among others, lung, esophagus, larynx, and colon cancer) and death from cancer worldwide. However, the effect of cigarette smoking in IBD course is surprising. Available data suggest that current smokers are more protected against UC, while at the same time smoking increases the risk of CD in a dose-dependent manner. Moreover, UC activity in smokers is lower when compared to non-smokers. Smoking UC patients have lower flare-up and hospitalization rates, less often need oral steroids, immunomodulators and biologic agents and—what is even more important—have lower colectomy rates compared to non-smokers [18, 19]. Interestingly, primary sclerosing cholangitis is observed almost exclusively in non-smoking patients [20].

In contrary, smokers with CD have a risk of flare-up increased by more than 50 %, an increased need of steroids, immunosuppressants and biologic agents as well as lower quality of life compared to non-smokers [21, 22]. Furthermore, smoking is connected with more frequent intestinal penetrating complications and a higher risk of being operated during disease course [18]. Moreover, the harmful effect of smoking is more marked in women and in patients with ileal disease [23, 24].

Finally, smokers with UC note symptom exacerbation when they quit smoking and symptom relief when they start smoking again [25]. Furthermore, smoking patients with UC who stop smoking experience an increase in disease activity; also, the number of hospital admissions within the first few years following the cessation of smoking is higher [26]. In contrast, patients with CD who quit smoking display similar disease severity to non-smokers and better course than continuing smokers [27].

To conclude, in clinical practice patients with IBD should be encouraged to quit smoking to reduce the risk of lung cancer and cardiovascular disease. However, smokers with UC should be informed that quitting smoking may potentially increase disease activity. Therefore, doctors should intensify treatment in patients with UC who plan to stop smoking.

10.1.2.2 Alcohol Consumption

The role of alcohol in causing or aggravating IBD in still unclear [28]. Usual consumption of alcohol (defined as alcoholic drinks 1–4 days per week) reduces the risk of UC when compared with less frequent use (odds ratio = 0.57, 95 % CI: 0.37–0.86) [29]. In line, light alcoholic drinking has protective effect against development of UC. Nevertheless, this effect disappears when smoking is included [30].

10.1.2.3 Physical Activity

The role of exercise in IBD has not been well studied, while some older epidemiological data suggest that physical activity is associated with a decreased risk of CD [18]. Results from a study, in which the effect of one-hour exercise in a cycle ergometer in six males with ileal CD was evaluated suggest that a moderate-intensity exercise program is probably safe for those patients [31]. In another study Loudon et al. [32] investigated the effects of a three-month low-intensity group walking program in CD patients. Significant improvement in CD activity and IBD Questionnaire were observed. In line, Ng et al. [33] using similar three months program found an improvement in quality of life in CD patients, with no exacerbation of disease symptoms.

Nowadays, the role of physical activity and exercise in the prevention and treatment of bone loss is well established in healthy population [35]. Furthermore, patients with CD in remission have decreased muscle function, which is self-described as a reduced strength and endurance, especially for lower limbs [34]; the may lead in longer term to osteoporosis. Thus, despite the fact that there is no clear evidence that exercise has any effect on IBD itself, some low-intensity training should be implemented to improve muscular mass and to prevent osteoporosis in IBD patients. A study performed by Robinson et al. [36] provides evidence supporting this recommendation, showing that low-intensity exercise in CD patients significantly improves bone mineral density after one year. Moreover, the increases in bone mineral density were correlated with the number of exercise sessions completed [36].

10.1.2.4 Obesity

While previously obesity in IBD has been considered unusual, nowadays its prevalence is increasing in CD, simultaneously with an increased prevalence in the whole population [37]. For example, epidemiological data from Scotland showed a prevalence of 18 % of obesity in CD [38].

Considering disease severity, overweight or obese patients with UC had to undergo colectomy more often than normal weight patients [38]. Interestingly, patients with CD had lower levels of surgery in the obese group compared to the normal-weight group [38]. On the other hand, other studies showed that obese CD patients had shorter time to first surgery [37, 39]. Taken together, it can be suggested that obese patients with IBD may require more aggressive medical therapy with avoidance of corticosteroids and should be encouraged to lower their body weight.

10.2 Intestinal Microflora

The human GI tract is colonized by a wide variety of microorganisms. Interestingly, more than 70 % of all microbes in the human body are in the colon, where they constitute a relatively stable ecosystem. Instantly after birth, oral cavity and gut are settled by an extensive range of microbes, mainly bacteria. Microbial load in the GI tract is not homogenous and ranges from 10^1 to 10^3 in the stomach and duodenum, progressing through the ileum in order to achieve a total number of 10^{11}–10^{12} bacterial cells per gram in the colon [40].

Commonly called human microbiota, it consists of trillions of organisms from over 1000 species [41]. The most abundant species are members of the phyla *Bacterioides* and *Firmicutes*. Bacterial diversity changes with age [42]. Apart from *Bacterioides* and *Firmicutes* phyla, adult gut ecosystem is colonized by *Actinobacteria*, *Proteobacteria* and *Verruvomicrobia* [43, 44]. It is important to emphasize that—among anaerobic bacteria in the human colon—*Bifidobacterium* spp and *Faecalibacterium* spp. can be found. Moreover, oxygen tolerant bacteria, such as *Lactobacillus* spp. may also appear in the gut, but in low number. The population of aerobic bacteria is fluctuating over time and depends on diet and other environmental factors, such as hygiene, climate, geography and ethnicity [45].

The GI tract microbiota is known to play an important role in the regulation of metabolic functions and maintenance of immune homeostasis. Microbiota is responsible for performing important biochemical reactions for host physiology, including degradation of xenobiotic substances, vitamin biosynthesis, fermentation of indigestible polysaccharides into beneficial short chain fatty acids, immune development and intestinal homeostasis maintenance [46].

Detrimental alterations in microbiota structure and functions causing the loss of ability to maintain homeostasis in GI tract are considered as a dysbiosis. Nowadays, a variety of pathologies are connected with the changes not only in the structure, but also the function of the gut microbiota, thereby suggesting a linkage between dysbiosis and disease etiology [46]. Namely, multiple disorders including type 2 diabetes, allergies, neurological disorders as well as IBD can be associated with dysbiosis [47].

IBD can be described as an immune-mediated disorder that originates from a breakdown of the normal symbiosis between the mucosal immune system and the commensal microbiota. This leads to the development of aberrant reactivity against intraluminal antigens and dysregulation of the innate and adaptive immunity. Consequently, these alterations may be responsible for tissue injury. Several studies have already demonstrated differences between microbiota of IBD patients and healthy individuals [48–53].

Regarding CD patients, lower diversity of microbiota, as compared with healthy individuals, was largely due to lower amounts of *Firmicutes*, especially *Clostridium leptum* phylogenetic group. Furthermore, another study reported that samples from small intestine of CD patients were less enriched in the *Bacillus* genus of *Firmicutes* and more rich in *Proteobacteria* [48]. Moreover, lower fecal concentrations of the *Bacterioides fragilis* group, *Clostridium coccoides* group, the *Atopobium* cluster and *Clostridium leptum* subgroup were observed in CD patients compared to healthy subjects [49].

Faecalibacterium prausnitzii, which is believed to be an important member of the normal gut flora, merits special attention. Research conducted by Sokol et al. [50] demonstrated a decrease in *F. prausnitzii* in samples from CD patients compared with healthy subjects. This is noteworthy in view of the fact that *F. prausnitzii*, through bacterial fermentation generate anti-inflammatory products which may act as a source of energy for the epithelial cells of intestine. Furthermore, they are believed to have impact on epithelial barrier integrity and play a role in intestinal immunomodulation [48].

Finally, higher abundance of *Enterobacteriaceae* has been observed in the intestinal samples of patients with CD. The increase in these bacteria may be due to intestinal inflammation itself, which promotes the growth of this strain of bacteria. Particularly, adherent-invasive *E. coli* (AIEC) strains were observed in higher abundance in mucosal samples from CD patients compared to healthy individuals [48].

The reduction of bacterial diversity and richness was also observed in UC patients. It included changes in abundance of *Firmicutes*, especially *Clostridium leptum* clusters, *Fecalibacterium prausnitzii*, *Clostridium coccoides*, *Roseburia*, *Ruminococcus*, *Enterococcus* and *Lactobacillus* [51]. Moreover, increased *Proteobacteria* populations, such as *Eschericia* sp., *Helicobacter* sp. and *Campylobacter* sp. were also reported [51]. Similarly to CD patients, AIEC has been observed in UC patients and implicated in the pathogenesis of UC [52]. Interestingly, AIEC were isolated from stools and rectal biopsies of UC patients during relapses and remissions.

It is worth mentioning that *Campylobacter* sp. is significantly more frequently detected in UC patients compared to controls [51]. These findings suggest that in UC patients specific immunological defect appears, which results in an inability to eliminate *Campylobacter* spp. *Fusobacterium varium* is another species of commensal bacteria increased in inflamed mucosa of UC patients. *F. varium* is believed to be responsible for production of high concentrations of butyric acid, which causes intestinal lesions in mice similar to those observed in human UC patients [53]. In contrast, many reports have demonstrated a decrease in *Faecalibacterium prausnitzii* in UC patients. Other studies showed that patients with UC had lower numbers of *Butyricicoccus* bacteria in their stools as well as lower abundance of *Roseburia hominis*. These three bacteria belong to *Clostridium leptum* group within the *Firmicutes* family and are known to produce butyrate [51].

It is indisputable that microbiota dysbiosis increases pathogenic and pro-inflammatory bacteria and decreases beneficial and anti-inflammatory bacteria. Genetic susceptibility of IBD patients leads to defective mucosal barrier function, which promotes invasion of pathogenic bacteria causing ulcerations and inflammation of the mucosa. Subsequent studies focused on the linkage between microbiota and the mucosal immune system are essential to understand the pathogenesis of IBD.

10.3 Psychological Aspects and Treatment

10.3.1 Introduction

Crohn's disease and ulcerative colitis are chronic disorders characterized by the presence of unpleasant symptoms from the GI tract impairing everyday functioning. IBD is often accompanied by harmful extraintestinal manifestations, with unpredictable course resulting in a significant reduction in the quality of life (QOL). The QOL is lower in women and CD patients than in men and UC patients, respectively. Higher levels of depression and anxiety are observed in CD compared to UC patients. Furthermore, in CD patients QOL was associated with disease activity; lower QOL was observed in relapse in comparison to patients with remission. Interestingly, patients after biological therapy with infliximab who achieved remission had higher quality of life than patients who did not received remission. In turn, in the UC patients the quality of life was connected with the extension of colon inflammation. Generally, patients after surgery presented lower QOL compared to patients treated pharmacologically.

Inflammatory bowel disease affects mainly young people, at the life phase when they obtain proper education and experience. Interestingly, research have suggested that patients with IBD have higher level of education comparing to healthy individuals. In contrast, patients with IBD have higher risk of not finishing higher school. Currently, it seems that besides difficulties with attendance in class patients with IBD achieve comparable level of education to healthy individuals. Furthermore, process of education among patients with IBD may be extended. Summarizing, IBD itself, as well as accompanying symptoms may affect future plans of patients and the type of obtained education.

10.3.2 Psychological Symptoms in IBD

There is growing evidence that psychological factors play a role in the pathophysiology and the course of IBD. Presence of stressful events, as well as perception of stress is believed to contribute—along with other factors such as the use of NSAIDs or nfections—to triggering flares of IBD [54]. Family stress was the most commonly reported. Patients with IBD also report stress accompanied with work, school and finances to be responsible for exacerbations of IBD.

The levels of depression and anxiety in patients with IBD are reported to be higher versus general population, but lower than in patients with functional bowel disorders [55]. Depressive disorder appears to affect more commonly older people and individuals with previous history of a psychiatric disorder. Moreover, females may have a higher risk of disease activity and relapses than men [56]. It has been suggested that depression, anxiety and impaired quality of life may exhibit negative influence on the course of IBD. Moreover, patients with IBD take more medications, such as antidepressants and anxiolytics than the healthy populations. Use of antidepressants to treat depression in IBD patients was found to be associated with decrease in relapse rates and steroid use [55].

Both depression and anxiety precede UC significantly more often, whereas no such relationship was seen in CD [55]. The association with UC is strongest when depression and anxiety are diagnosed in the same year as UC. Nevertheless, the origin of depression and anxiety in IBD patients remains not completely explained. It is suggested that depression and anxiety are consequences of IBD symptoms, such as frequent stools, abdominal pain and bloating.

10.3.3 Psychiatric Therapy in IBD

Anxiety and depression are highly treatable disorders with several agents with proven efficacy. Most efficient are selective serotonin reuptake inhibitors (SSRIs) and serotonin norepinephrine reuptake inhibitors (SNRI). To name only few citalopram, escitalopram, fluoxetine, paroxetine, sertraline and venlafaxine.

Gastroenterologists reported that antidepressants were successful in reducing pain, gut irritability and urgency of defecation in IBD patients. Interestingly, approximately 29 % of IBD patients had used an antidepressant at some time of their life. The study conducted by Goodhand et al. [57] showed that patients taking antidepressants reported fewer relapses and steroid use in the year after starting treatment than in the year before.

A systemic review of SSRIs indicated that although the medications were similar in efficacy, there were differences in their side effect profiles [55]. Gastrointestinal side effects have been reported for antidepressant agents and can be a crucial concern to the IBD patients. Despite the fact that these side effects are generally dose related and tend to decrease or even vanish over the first weeks of treatment, they can reduce patient's adherence to medications and convince them to discontinue therapy. The most common side effects are nausea and vomiting, followed by diarrhea, drowsiness, anxiety, headache, insomnia, fatigue. Considering SSRI medications, weight gain becomes a major problem. Furthermore, a decreased sexual functioning which is a relatively common side effect of antidepressant agents may be a concern for IBD patients and impose treatment discontinuation.

10.4 Cooperation Between the Patient and the Doctor

- Despite the fact that the role of a doctor in medical care of patients with IBD depends on clinical situation, the major goal is to satisfy multidisciplinary needs of this group.
- Regarding complexity of IBD in the treatment process, primary care physicians and specialists such as a gastroenterologist, a surgeon as well as a psychologist and dietitian should be involved.
- IBD is chronic disease characterized by unforeseeable episodes of remissions and relapses [58], hence patients usually need long-lasting and accurate medical attention, which includes regular appointments and even phone calls. During these appointments, doctors should assess the development of disease symptoms and provide necessary support to all patients. Therefore building relations based on trust and good contact with a patient is crucial.
- During the treatment process, the doctor should bear in mind that UC as well as CD are chronic diseases, seriously affecting not only the GI tract, but also general functioning of the whole organism.
- The main goal of the treatment, which should be adjusted individually to each and every patient, is to induce and maintain remission with minimal adverse effect of used drugs and to improve patient quality of life through limitation of inflammation in GI tract as well as elimination of symptoms such abdominal pain, diarrhea and bleeding from anus.
- Given the complexity of IBD, the fundamental role of the doctor is to supervise and educate patient about disease process, used drugs, clinical examinations and laboratory tests as well as proper diet and lifestyle.

Patients with IBD may present a whole range of different symptoms depending on disease localization and severity. During the course of IBD, especially UC, extraintestinal manifestations may occur in addition to GI symptoms such as uveitis, arthritis or primary sclerosing cholangitis [58]. For this reason, to provide proper medical care the doctor should obtain detailed information about the symptoms, complications and extraintestinal manifestations. In particular, the doctor should pay attention to presence, severity and localization of pain. Moreover, during appointments the doctor should obtain information about frequency of defecations, pain during defecation, abnormal stool, blood in stool and presence of symptoms such as vomiting, nausea, dyspepsia, flatulence. Obtaining information about symptoms will help the doctor to objectively decide if the patient is in remission or relapse.

During the course of IBD, several complications such as fistulas, abscesses, strictures and even bowel obstruction may occur, especially in CD patients. UC patients have greater risk of megacolon toxicum and colorectar cancer [58, 59]. Hence, during appointment the doctor should pay particular attention to every symptoms suggesting development of IBD complications.

To improve relation with the patient and to strengthen patient's sense of self-reliance, the doctor should ask patient to describe by own words their feelings about disease, and the symptoms which make them anxious. Unraveling patient's concerns will help the doctor choose the best therapeutic option for the patient.

Considering that IBD has chronic and unpredictable course, it is essential to educate patients and theirs family about disease in clear and comprehensible way. IBD may affect people with different education levels, hence during appointments the doctor should assess patient's level of understanding of the disease and identify and remove all obstacles in educational process. Majority of patients are treated ambulatory. Therefore, to increase efficacy of the education process additional materials, such as books, booklets and posters should be used as well as information about support groups should be provided. Meeting with other patients should also be beneficial for patients with IBD. It may help them realize that - despite episodes of relapses and remissions—patients with IBD lead normal and active life with low intensity of symptoms. Showing successful life of other IBD patients may be the source of hope and encouragement to follow doctor's instructions.

Knowing the localization of the disease allows to explaining the nature of the disease and symptoms and plan proper treatment, which will be most efficient in this particular patient. It is worth mentioning that the doctor's responsibility is to inform patients that certain habits may have impact on the course of the disease. Studies demonstrated that smoking have negative effect on the course of CD, thus patients should be encouraged to quit smoking. Moreover, use of NSAIDs may exacerbate IBD symptoms and patients should limit the use of these drugs.

Educating the patient is a long process, which should be carefully conducted by the doctor to avoid overwhelming of the patient. Short education sessions and regular appointments should be considered as most preferable solutions. Additional materials such as posters and booklets developed by support groups and organizations dedicated to IBD should be available to all patients [60]. Many patients may benefit from participation in support groups. Patients often find it easier to talk about their fears and symptoms with people with similar experience. Patients who have contact with other people with similar experience have an increased psychological comfort in comparison to people not attending support groups meetings. [61]. Nowadays, many patients obtain information about IBD from the world wide web. Those patients should be instructed to use only trustworthy addresses, provided by their doctors or support groups. Information from internet forums and discussion groups may mislead patients and result in the loss of trust in the doctor—patient relation

Nowadays, several drugs such as aminosalicylates, corticosteroids, immunomodulators and biological agents are available in IBD therapy. However, IBD treatment is not deprived of side effects. Considering importance of proper drug use and their side effects, it is crucial to educate patients in this issue during ambulatory visits and hospitalizations. It is necessary to explain the type of the drug, administration route and dose as well as its therapeutic role and side effects. Increasing patient's knowledge about drugs may lead to better commitment in therapeutic process. Moreover, patient's knowledge about symptoms of drug intoxication, such as abdominal pain, nausea, paresthesia and muscle weakness is essential. Patient with that knowledge will be able to discontinue treatment and rapidly contact the doctor. Additionally, the doctor should inform the patient about other not life-threating adverse effects which may perturb the patient, such as change in urine color after sulfasalazine use and many others.

References

1. Carbonnel F, Jantchou P, Monnet E, Cosnes J (2009) Environmental risk factors in Crohn's disease and ulcerative colitis: an update. Gastroenterol Clin Biol 33(Suppl 3):S145–S157
2. Molodecky NA, Soon IS, Rabi DM, Ghali WA, Ferris M, Chernoff G et al (2012) Increasing incidence and prevalence of the inflammatory bowel diseases with time, based on systematic review. Gastroenterology 142(1):46–54
3. Bernstein CN (2010) New insights into IBD epidemiology: Are there any lessons for treatment? Dig Dis 28(3):406–410
4. Kappelman MD, Rifas-Shiman SL, Kleinman K, Ollendorf D, Bousvaros A, Grand RJ et al (2007) The prevalence and geographic distribution of Crohn's disease and ulcerative colitis in the United States. Clin Gastroenterol Hepatol 5(12):1424–1429
5. Hou JK, Abraham B, El-Serag H (2011) Dietary intake and risk of developing inflammatory bowel disease: a systematic review of the literature. Am J Gastroenterol 106(4):563–573
6. Dixon LJ, Kabi A, Nickerson KP, McDonald C (2015) Combinatorial effects of diet and genetics on inflammatory bowel disease pathogenesis. Inflamm Bowel Dis 21(4):912–922
7. Russel MG, Engels LG, Muris JW, Limonard CB, Volovics A, Brummer RJ et al (1998) Modern life' in the epidemiology of inflammatory bowel disease: a case-control study with special emphasis on nutritional factors. Eur J Gastroenterol Hepatol 10(3):243–249
8. Sakamoto N, Kono S, Wakai K, Fukuda Y, Satomi M, Shimoyama T, et al. (2005) Dietary risk factors for inflammatory bowel disease: a multicenter case-control study in Japan. Inflamm Bowel Dis 11(2)

9. Reif S, Klein I, Lubin F, Farbstein M, Hallak A, Gilat T (1997) Pre-illness dietary factors in inflammatory bowel disease. Gut 40(6):754–760
10. Ananthakrishnan AN, Khalili H, Konijeti GG, Higuchi LM, de Silva P, Fuchs CS et al (2014) Long-term Intake of dietary fat and risk of ulcerative colitis and Crohn's disease. Gut 63 (5):776–784
11. Ananthakrishnan AN, Khalili H, Konijeti GG, Higuchi LM, de Silva P, Korzenik JR et al (2013) A prospective study of long-term intake of dietary fiber and risk of Crohn's disease and ulcerative colitis. Gastroenterology 145(5):970–977
12. Owczarek D, Rodacki T, Domagala-Rodacka R, Cibor D, Mach T (2016) Diet and nutritional factors in inflammatory bowel diseases. World J Gastroenterol 22(3):895–905
13. Cohen AB, Lee D, Long MD et al (2013) Dietary patterns and self-reported associations of diet with symptoms of inflammatory bowel disease. Dig Dis Sci 58(5):1322–1328. doi:10.1007/s10620-012-2373-3
14. Herfarth HH, Martin CF, Sandler RS, Kappelman MD, Long MD (2014) Prevalence of a gluten free diet and improvement of clinical symptoms in patients with inflammatory bowel diseases. Inflamm Bowel Dis 20(7):1194–1197
15. Marlow G, Ellett S, Ferguson IR, Zhu S, Karunasinghe N, Jesuthasan AC et al (2013) Transcriptomics to study the effect of a Mediterranean-inspired diet on inflammation in Crohn's disease patients. Human Genomics 7(1):1–9
16. Cohen SA, Gold BD, Oliva S, Lewis J, Stallworth A, Koch B, et al. Clinical and mucosal improvement with specific carbohydrate diet in pediatric crohn disease. J Pediatr Gastroenterol Nutrition 59(4)
17. Gearry RB, Irving PM, Barrett JS, Nathan DM, Shepherd SJ, Gibson PR (2009) Reduction of dietary poorly absorbed short-chain carbohydrates (FODMAPs) improves abdominal symptoms in patients with inflammatory bowel disease—A pilot study. J Crohn's Colitis 3 (1):8–14
18. Cosnes J (2010) Smoking, physical activity, nutrition and lifestyle: environmental factors and their impact on IBD. Dig Dis 28(3):411–417
19. Lakatos PL, Szamosi T, Lakatos L (2007) Smoking in inflammatory bowel diseases: good, bad or ugly? World J Gastroenterol 13(46):6134–6139
20. Loftus EV Jr, Sandborn WJ, Tremaine WJ, Mahoney DW, Zinsmeister AR, Offord KP et al (1996) Primary sclerosing cholangitis is associated with nonsmoking: a case- control study. Gastroenterology 110(5):1496–1502
21. Cosnes J, Carbonnel F, Carrat F, Beaugerie L, Cattan S, Gendre J (1999) Effects of current and former cigarette smoking on the clinical course of Crohn's disease. Aliment Pharmacol Ther 13(11):1403–1411
22. Russel MGVM, Volovics A, Schoon EJ, van Wijlick EHJ, Logan RF, Shivananda S, et al (1998) Inflammatory bowel disease: is there any relation between smoking status and disease presentation? Inflamm Bowel Dis 4(3)
23. Persson PG, Ahlbom A, Hellers G (1990) Inflammatory bowel disease and tobacco smoke—A case-control study. Gut 31(12):1377–1381
24. Cosnes J (2004) Tobacco and IBD: relevance in the understanding of disease mechanisms and clinical practice. Best Pract Res Clin Gastroenterol 18(3):481–496
25. Motley RJ, Rhodes J, Ford GA, Wilkinson SP, Chesner IM, Asquith P et al (1987) Time relationships between cessation of smoking and onset of ulcerative colitis. Digestion 37 (2):125–127
26. Beaugerie L, Massot N, Carbonnel F, Cattan S, Gendre JP, Cosnes J (2001) Impact of cessation of smoking on the course of ulcerative colitis. Am J Gastroenterol 96(7):2113–2116
27. Cosnes J, Beaugerie L, Carbonnel F, Gendre J (2001) Smoking cessation and the course of Crohn's disease: an intervention study. Gastroenterology 120(5):1093–1099
28. El-Tawil AM (2013) Epidemiology and inflammatory bowel diseases. World J Gastroenterol 19(10):1505–1507
29. Nakarnura Y, Labarthe DR (1994) A case-control study of ulcerative colitis with relation to smoking habits and alcohol consumption in Japan. Am J Epidemiol 140(10):902–911

30. Jiang L, Xia B, Li J, Ye M, Deng C, Ding Y, et al (2007) Risk factors for ulcerative colitis in a chinese population: an age-matched and sex-matched case-control study. J Clin Gastroenterol 41(3)
31. D'Inca R, Varnier M, Mestriner C, Martines D, D'Odorico A, Sturniolo GC (1999) Effect of moderate exercise on Crohn's disease patients in remission. Ital J Gastroenterol Hepatol 31 (3):205–210
32. Loudon CP, Corroll V, Butcher J, Rawsthorne P, Bernstein CN (1999) The effects of physical exercise on patients with Crohn's disease. Am J Gastroenterol 94(3):697–703
33. Ng V, Millard W, Lebrun C, Howard J (2007) Low-intensity exercise improves quality of life in patients with Crohn's disease. Clin J Sport Med 17(5):384–388
34. Wiroth JB, Filippi J, Schneider SM, Al-Jaouni R, Horvais N, Gavarry O, et al (2005) Muscle Performance in patients with Crohn's disease in clinical remission. Inflammatory Bowel Dis 11(3)
35. Lee N, Radford-Smith G, Taaffe DR (2005) Bone loss in Crohn's disease: exercise as a potential countermeasure. Inflamm Bowel Dis 11(12):1108–1118
36. Robinson RJ, Krzywicki T, Almond L, Al–Azzawi F, Abrams K, Iqbal SJ et al (1998) Effect of a low-impact exercise program on bone mineral density in Crohn's disease: a randomized controlled trial. Gastroenterology 115(1):36–41
37. Blain A, Cattan S, Beaugerie L, Carbonnel F, Gendre JP, Cosnes J (2002) Crohn's disease clinical course and severity in obese patients. Clin Nutr 21(1):51–57
38. Steed H, Walsh S, Reynolds N (2009) A brief report of the epidemiology of obesity in the inflammatory bowel disease population of tayside, Scotland. Obes Facts 2(6):370–372
39. Hass DJ, Brensinger CM, Lewis JD, Lichtenstein GR () The impact of increased body mass index on the clinical course of Crohn's disease. Clin Gastroenterol Hepatol 4(4):482–488
40. Berg RD (1996) The indigenous gastrointestinal microflora. Trends Microbiol 4(11):430–435
41. Zatorski H, Fichna J (2014) What is the future of the gut microbiota-related treatment? Toward modulation of microbiota in preventive and therapeutic medicine. Front Med (Lausanne) 1:19
42. Yatsunenko T, Rey FE, Manary MJ, Trehan I, Dominguez-Bello MG, Contreras M et al (2012) Human gut microbiome viewed across age and geography. Nature 486(7402):222–227
43. Arumugam M, Raes J, Pelletier E, Le Paslier D, Yamada T, Mende DR et al (2011) Enterotypes of the human gut microbiome. Nature 473(7346):174–180
44. Salonen A, Salojärvi J, Lahti L, de Vos WM (2012) The adult intestinal core microbiota is determined by analysis depth and health status. Clin Microbiol Infect 18:16–20
45. Marques TM, Cryan JF, Shanahan F, Fitzgerald GF, Ross RP, Dinan TG et al (2014) Gut microbiota modulation and implications for host health: dietary strategies to influence the gut–brain axis. Innovative Food Sci Emerg Technol 22:239–247
46. Buttó LF, Haller D (2016) Dysbiosis in intestinal inflammation: Cause or consequence. Int J Med Microbiol 306(5):302–309
47. Schippa S, Conte MP (2014) Dysbiotic events in gut microbiota: impact on human health. Nutrients 6(12):5786–5805
48. Øyri SF, Müzes G, Sipos F (2015) Dysbiotic gut microbiome: a key element of Crohn's disease. Comp Immunol Microbiol Infect Dis 43:36–49
49. Takaishi H, Matsuki T, Nakazawa A, Takada T, Kado S, Asahara T et al (2008) Imbalance in intestinal microflora constitution could be involved in the pathogenesis of inflammatory bowel disease. Int J Med Microbiol 298(5):463–472
50. Sokol H, Seksik P, Furet JP, Firmesse O, Nion-Larmurier I, Beaugerie L, et al (2009) Low counts of Faecalibacterium prausnitziiin colitis microbiota. Inflamm Bowel Dis 15(8)
51. Ohkusa T, Koido S (2015) Intestinal microbiota and ulcerative colitis. J Infect Chemother 21 (11):761–768
52. Burke DA, Axon AT (1988) Adhesive Escherichia coli in inflammatory bowel disease and infective diarrhoea. BMJ 297(6641):102–104

53. Ohkusa T, Okayasu I, Ogihara T, Morita K, Ogawa M, Sato N (2003) Induction of experimental ulcerative colitis by Fusobacterium varium isolated from colonic mucosa of patients with ulcerative colitis. Gut 52(1):79–83
54. Bernstein CN, Singh S, Graff LA, Walker JR, Miller N, Cheang M (2010) A prospective population-based study of triggers of symptomatic flares in IBD. Am J Gastroenterol 105 (9):1994–2002
55. Filipović BR, Filipović BF (2014) Psychiatric comorbidity in the treatment of patients with inflammatory bowel disease. World J Gastroenterol 20(13):3552–3563
56. Filipović BR, Filipović BF, Kerkez M, Milinić N, Randelovć T (2007) Depression and anxiety levels in therapy-naive patients with inflammatory bowel disease and cancer of the colon. World J Gastroenterol 13(3):438–443
57. Goodhand JR, Greig FIS, Koodun Y, McDermott A, Wahed M, Langmead L, et al (2012) Do Antidepressants influence the disease course in inflammatory bowel disease? A retrospective case-matched observational study. Inflamm Bowel Dis 18(7)
58. Sobczak M, Fabisiak A, Murawska N, Wesolowska E, Wierzbicka P, Wlazlowski M et al (2014) Current overview of extrinsic and intrinsic factors in etiology and progression of inflammatory bowel diseases. Pharmacol Rep 66(5):766–775
59. Sobczak M, Wlazłowski M, Zatorski H, Sałaga M, Fichna J (2014) Current overview of colitis-associated colorectal cancer. Central Eur J Biol 9(11):1022–1029
60. Husain A, Triadafilopoulos G (2004) Communicating with patients with inflammatory bowel disease. Inflamm Bowel Dis 10(4)
61. Leshem RN (2003) Inflammatory bowel disease support groups: a primer for gastroenterology nurses. Gastroenterol Nurs 26(6)

Correction to: Pharmacological and Clinical Treatment of Irritable Bowel Syndrome

Maciej Sałaga and Paula Mosińska

Correction to:
Chapter 4 in: J. Fichna (ed.), *Introduction to Gastrointestinal Diseases Vol. 1*, https://doi.org/10.1007/978-3-319-49016-8_4

The original version of the chapter "Pharmacological and Clinical Treatment of Irritable Bowel Syndrome" was inadvertently published with spell errors in Fig. 4.1 in page 45 and in-line text in page 47. The errors have now been corrected in the chapter.

The updated version of this chapter can be found at
https://doi.org/10.1007/978-3-319-49016-8_4

J. Fichna (ed.), *Introduction to gastrointestinal diseases Vol. 1*,
https://doi.org/10.1007/978-3-319-49016-8_11

Summary

We often tend to forget how important is the gastrointestinal tract for the maintenance of systemic homeostasis, and how debilitating its malfunctions may be for both, physical and mental condition. The book should give a better understanding on what triggers functional and inflammatory diseases, how they could be diagnosed and what treatments are available. On the one hand, the book should encourage the patient to seek professional advice when symptoms occur, but also make them more responsible for their health through proper diet and healthy living. On the other hand, the book should also help the doctor take proper care of the patient, who is often confused and strained not only by disease symptoms, but also by the need to seek for medical advice.

Our knowledge on functional and inflammatory diseases expands rapidly. With this book we get the patient and the doctor acquainted with most crucial information, but also we encourage a new, more personalized approach to the patient, along with efficient teaming between the patient and the doctor.

J. Fichna (ed.), *Introduction to gastrointestinal diseases Vol. 1*,
DOI 10.1007/978-3-319-49016-8

MIX
Papier aus verantwortungsvollen Quellen
Paper from responsible sources
FSC® C105338

If you have any concerns about our products,
you can contact us on
ProductSafety@springernature.com

In case Publisher is established outside the EU,
the EU authorized representative is:
Springer Nature Customer Service Center GmbH
Europaplatz 3, 69115 Heidelberg, Germany

Printed by Libri Plureos GmbH
in Hamburg, Germany